S0-CAH-241

Synthesis of Carbohydrates through Biotechnology

ACS SYMPOSIUM SERIES **873**

Synthesis of Carbohydrates through Biotechnology

Peng George Wang, Editor
Wayne State University

Yoshi Ichikawa, Editor
Optimer Pharmaceuticals, Inc.

Sponsored by the
ACS Divisions of Carbohydrate Chemistry

American Chemical Society, Washington, DC

Library of Congress Cataloging-in-Publication Data

Synthesis of carbohydrates through biotechnology / Peng George Wang, editor, Yoshi Ichikawa, editor ; sponsored by the ACS Division of Carbohydrate Chemistry.

p. cm.—(ACS symposium series ; 873)

Includes bibliographical references and index.

ISBN 0-8412-3865-0

1. Carbohydrates—Congresses. 2. Organic compounds—Synthesis—Congresses.

I. Wang, Peng George. II. Ichikawa, Yoshi. III. American Chemical Society. Division of Carbohydrate Chemistry. IV. American Chemical Society. Meeting (224th : 2002 : Boston, Mass.) V. Series.

QD320.S92 2004
547′.780459—dc22 2003066455

The paper used in this publication meets the minimum requirements of American National Standard for Information Sciences—Permanence of Paper for Printed Library Materials, ANSI Z39.48–1984.

Distributed by Oxford University Press

PRINTED IN THE UNITED STATES OF AMERICA

Foreword

The ACS Symposium Series was first published in 1974 to provide a mechanism for publishing symposia quickly in book form. The purpose of the series is to publish timely, comprehensive books developed from ACS sponsored symposia based on current scientific research. Occasionally, books are developed from symposia sponsored by other organizations when the topic is of keen interest to the chemistry audience.

Before agreeing to publish a book, the proposed table of contents is reviewed for appropriate and comprehensive coverage and for interest to the audience. Some papers may be excluded to better focus the book; others may be added to provide comprehensiveness. When appropriate, overview or introductory chapters are added. Drafts of chapters are peer-reviewed prior to final acceptance or rejection, and manuscripts are prepared in camera-ready format.

As a rule, only original research papers and original review papers are included in the volumes. Verbatim reproductions of previously published papers are not accepted.

ACS Books Department

Contents

Preface ix

1. **Approaches to the Enzymatic Carbohydrate Synthesis** 1
   Przemyslaw Kowal, Jun Shao, Mei Li, Wen Yi, Hanfen Li, and Peng George Wang

2. **Exploring Carbohydrate Diversity through Automated Synthesis** 11
   Obadiah J. Plante

3. **Glycosynthesis for Drug Discovery: Globo H Synthesis by OPopS** 23
   Shirley A. Wacowich-Sgarbi, David Rabuka, Paulo W. M. Sgarbi, and Yoshi Ichikawa

4. **Probing the Antigenic Diversity of Sugar Chains** 39
   Ruby Wang, Brian J. Trummer, Ellen Gluzman, Chao Deng, and Denong Wang

5. **Chemoenzymatic Synthesis of Lactosamine and (α2→3)Sialylated Lactosamine Building Blocks** 53
   Fengyang Yan, Seema Mehta, Eva Eichler, Warren Wakarchuk, and Dennis M. Whitfield

6. **Synthesis of Bioactive Glycopeptides through Endoglycosidase-Catalyzed Transglycosylation** 73
   Lai-Xi Wang*, Suddham Singh, and Jiahong Ni

7. **Strategies for Synthesis of an Oligosaccharide Library Using a Chemoenzymatic Approach** 93
   Ola Blixt and Nahid Razi

8. **Artificial Golgi Apparatus: Direct Monitoring of Glycosylation Reactions on Automated Glycosynthesizer** 113
   Shin-Ichiro Nishimura, Noriko Nagahori, Reiko Sadamoto, Kenji Monde, and Kenichi Niikura

9. **Sugar Engineering with Glycosaminoglycan Synthases** **125**
Paul L. DeAngelis

10. **Production of Oligosaccharides Using Engineered Bacteria: Engineering of Exopolysaccharides from Lactic Acid Bacteria** **139**
Laure Jolly, Véronique Tornare, and Sunil Kochhar

11. **Production of Oligosaccharides by Coupling Engineered Bacteria** **153**
Satoshi Koizumi

12. **Synthesis of Glycoconjugates through Biosynthesis Pathway Engineering** **165**
Mei Li, Jun Shao, Min Xiao, and Peng George Wang

Indexes

Author Index **185**

Subject Index **187**

Preface

At the cell surface, carbohydrates and their conjugates mediate a vast array of biological processes and confer specificity to a wide array of cellular processes such as immunological responses, molecular recognition, and host–pathogen interactions. Furthermore, carbohydrate structures often change upon carcinogenic transformation. Glycomics, which is the study of an organism's entire array of oligosaccharides, is now emerging as the third informatics wave after genomics and proteomics. It will be key to our full understanding of the molecular phenotype.

No useful information carrier exists for the glycomics, which is equivalent to DNA for genomics and proteomics, and the study of carbohydrate structure and activity is very difficult. Furthermore, the microheterogeneity associated with glycoforms of a glycoprotein makes such study more complicated.

Because of these complexities, the function of the majority of carbohydrates in biological systems remains unknown, despite recent advances in the structural elucidation of oligosaccharides. Consequently, carbohydrates are less interesting targets for drug development even though they are important components in many valuable therapeutics such as antibiotics and anticancer agents. We badly need reliable technologies to efficiently assemble glycoconjugates to enable their study. Therefore, we felt that it was a propitious time to assemble some of the leading chemists in glycoscience at this year's American Chemical Society (ACS) meeting to address this issue. We invited 11 leading scientists from academia and industry to present the state-of-the-art glycoscience technology and its implications for biotechnology—especially in drug discovery and development.

Because the title of our symposium was *Synthesis of Carbohydrates through Biotechnology,* we will discuss the most impor-tant factors in this area. Three major factors dictate the research activity in biotechnology: scientific merit (science), intellectual property (IP), and its application (market). How these factors intersect will identify the future opportunities in this field. Understanding these needs facilitates the scientific progress and gives a positive contribution to our society. We cannot only consider one area and isolate the searchers. Hopefully, this book will help you to identify what you can do in glycochemistry to contribute to the human health.

This book introduces current results and state-of-the-art methods in carbohydrate synthesis and production. The first chapter (Kowal et al.) provides an overall view on the large-scale production of oligosaccharides through enzymatic reactions and fermentation. Next we invited authors to write chapters to describe three most exciting technology developments in the field of carbohydrate science. The second chapter (Plante) describes the latest development in automated synthesis of oligosaccharides using Seeberger's "carbohydrate synthesizer" machine. The third chapter (Ichikawa and co-workers) give an exciting example on synthesizing an complex oligosaccharide Globo H using the "one-pot" synthesis technology developed in C. H. Wong's laboratory. In the fourth chapter, Denong Wang and co-workers provide an excellent account on their "glyco-chip" technology for probing the antigenic diversity of sugar chains. The next five chapters discuss the latest development on synthesis of carbohydrates using purified enzymes. The fifth chapter (Whitfield and co-workers) describes an excellent example on the combination of chemical and enzymatic methods for preparation of sialylated lactosamine building blocks. The sixth chapter (Wang et al.) provides an outstanding example on how to use endoglycosidase-catalyzed transglycosylation to prepare fairly complex *N*-glycan. In Chapter seven, Blixt and Razi give a comprehensive account on the enzymatic synthesis of fucosylated and sialylated galactosides–polylactosamines and gangliosides. In Chapter eight, Nishimura et al. describe their unique artificial Golgi Apparatus, which can be considered to be the biological version of the chemical "carbohydrate synthesizer". Moreover, in Chapter 9, DeAngelis focuses on what is considered the best example on the strength of enzymatic carbohydrate synthesis: preparation of complex glycosaminoglycan polymers. The last three chapters of this book are on the large-scale production of carbohydrates through fermentation. In Chapter 10, Jolly et al. demonstrate production of oligosaccharides using lactic acid bacteria. The eleventh chapter (Koizumi) focuses on the technology developed in Kyowa Hakko Kogyo Company to produce oligosac-charides by multiple engineered bacteria. The last chapter of this book (Wang and co-workers) describe the "superbug" approach on production of carbohydrates in P. G. Wang's laboratory.

We feel privileged to have attracted such a distinguished group of investigators and express our sincerest gratitude for their time and effort. We hope that this book will encourage more research efforts and investment on carbohydrate production and commercialization.

Acknowledgments

We thank the following companies and associations for contributing to the symposium: Kyowa Hakko Kogyo Company, Ltd., Neose Technologies Inc., ACS Corporation Associates, and the ACS Division of Carbohydrate Chemistry. The peer reviewers deserve special recognition for their conscientious review of the chapters.

Peng George Wang*
Department of Chemistry
Wayne State University
Detroit, MI 48202

Yoshi Ichikawa
Optimer Pharmaceuticals, Inc.
10110 Sorrento Valley Road
Suite C
San Diego, CA 92122

*Current affiliation: Departments of Chemistry and Biochemistry, The Ohio State University, 876 Biological Sciences Building, 484 West 12th Avenue, Columbus, OH 43210.

Chapter 1

Approaches to the Enzymatic Carbohydrate Synthesis

Przemyslaw Kowal[1], Jun Shao[1], Mei Li[1], Wen Yi[1], Hanfen Li[1], and Peng George Wang[1,2]

[1]Department of Chemistry, Wayne State University, Detroit, MI 48202
[2]Current address: Departments of Chemistry and Biochemistry, The Ohio State University, 876 Biological Sciences Building, 484 West 12th Avenue, Columbus, OH 43210

Polysaccharides are the least appreciated class of macromolecules though they constitute the largest part of the planet's bio-mass. This large bulk has perpetuated a compelling reason for their neglect: carbohydrates were thought to be biochemically uninteresting serving as structural and energy storage molecules. The second more insidious and perhaps contradictory reason for the relative slowness in carbohydrate research is that biochemically important carbohydrates are difficult to study. They seem haphazardly put together, without a template, creating enormous complexity that, at first, did not suggest specific functions. The past three decades have seen enormous strides in the understanding of the importance of carbohydrates feeding the quickly expanding and vibrant field of enzymatic carbohydrate synthesis. The clear goal of the research in this field is to supplant Nature as the supreme oligosaccharide maker to generate desired saccharides, be it natural or unnatural, in high yields, large quantities and at low cost.

Unlike the linear nature of the nucleic acid and protein molecules, polysaccharides exhibit highly branched structures and multiplicity of different, although related, types of linkages known as glycosidic bonds. Polysaccharide synthesis does not follow a linear template and the mechanisms that govern this synthesis are difficult to elucidate. Enzymes that catalyze the formation of glycosidic bonds in nature are referred to as glycosyltransferases (GTs). These enzymes use activated sugar molecules (donors) to attach saccharide residues to a variety of substrates (acceptors). The donor molecules most often are nucleotide diphospho sugars (e.g. UDP-Glc) but can also be sugar phosphates or disaccharides. The most common acceptor molecules are growing carbohydrate chains, however, a variety of other compounds such as lipids, proteins, or steroids can also serve this function [1]. The majority of carbohydrates present in cells occurs in the form of oligosaccharides that are attached to proteins or lipids which are collectively called glycoconjugates [2].

Oligosaccharides do not have a single specific function, however, it is believed that their major role is to serve as recognition markers. Oligosaccharide decorated proteins and lipids are fundamental to many important biological processes including fertilization, immune defense, viral replication, parasitic infection, cell growth, cell-cell adhesion, degradation of blood clots, inflammation and cancer metastasis [2-5]. Oligosaccharide structures change dramatically during development and unique sets of oligosaccharides are expressed at distinct stages of differentiation [6]. Oligosaccharides can modify the intrinsic properties of proteins to which they are attached by altering the stability, protease resistance, or quaternary structure. In addition, their bulk may occlude functionally important areas of proteins to modulate protein-ligand interactions and to affect the rate of processes which involve conformational changes.

Glycosylation is both cell type and tissue type dependent. It is also highly sensitive to alterations in cellular state. It offers the cell the ability to quickly adjust to changing external conditions by altering glycosylation patterns and by using glycosylation to fine tune the activities of receptors and enzymes that is not possible or too slow to be effective at the genetic level. Aberrant glycosylation is diagnostic of a number of diseases including rheumatoid arthritis and cancer.

Eukaryotic proteins destined for glycosylation are modified early after their synthesis in the endoplasmic reticulum (ER). The process is continued into the early Golgy where trimming and additional changes to the glycoforms take place. Carbohydrates are attached covalently through either a nitrogen atom in asparagine side chain (N-glycosylation) or oxygen atoms in the side chains of threonine and serine (O-glycosylation). N-glycosylation is protein sequence specific though not all possible sites are glycosylated suggesting that the 3D structure of the protein plays a role in determining the extent and type of its own glycosylation. A series of membrane-bound glycosidases and glycosyltransferases act sequentially on the growing oligosaccharide as it moves through the lumen of the ER and Golgi apparatus. Many different enzymes are involved in the processing pathways and some steps may be bypassed giving rise to a large number of glycoform variants of the same polypeptide. For example, the PrP^{C} human prion protein that has seen much scrutiny is known to possess 52 different glycans [7].

Glycolipids are sugar-containing lipids and are a major class of glycoconjugates synthesized by eukaryotic and prokaryotic cells. In general they consist of sphingosine (or one of its derivatives), one molecule of a fatty acid and carbohydrate residues of varying complexity as polar head groups. The amino group of the sphingoid base is attached to a long-chain fatty acyl group forming ceramide. The ceramide moiety anchors the glycolipid in the membrane, extending the saccharide head group out into the extracellular space. This localization has positioned glycolipids directly at the interface between the cell and its external environment, which may include many hostile organisms. Many pathogens or the toxins they produce must breach the cellular membrane in order to be infectious. For example, *Escherichia coli* O157, the so-called "hamburger bug", is responsible for serious, and sometimes deadly, outbreaks of food poisoning [8-10]. As part of its infection strategy, *E. coli* O157 produces a toxin (verotoxin) that binds to the carbohydrate part of the glycolipid globotriosylceramide (Galα1,4Galβ1,4Glcβ1Cer) in the membranes of many cells.

Bacterial cells express an enormous variety of polysaccharide structures. These entities are usually found protruding from the outer membrane and include exopolysaccharide (EPS), lipopolysaccharide (LPS) and lipooligosaccharide (LOS), though protein glycosylation has also been found [11]. EPS can be found in both Gram positive and Gram negative cells and is made of repeating polysaccharide units. Bacteria commit significant resources to the synthesis of carbohydrates. For example, LPS accounts for about 2% of the cells dry weight.

The increased understanding of the functions carbohydrates play in cellular interactions have spurred a tremendous growth in glycobiology. Oligosaccharides are now widely recognized as potential pharmaceuticals for the prevention of infection, neutralization of toxins, cancer immunotherapy and even antiviral therapy. This, in turn, has fueled a large and as yet unfulfilled demand for carbohydrates. During the past few decades, a number of chemical approaches have been developed for the synthesis of oligosaccharides [12]. Chemical methods often require multiple protection-deprotection steps and long synthetic routes and have therefore been of limited importance to industrial and scientific communities [13]. Biocatalytic approaches employing enzymes or genetically engineered whole cells are powerful and complimentary alternatives to chemical methods alone [14].

Enzymes have been used extensively to simplify the synthesis of complex oligosaccharides and glycoconjugates. Glycosyltransferases and glycosidases are valuable catalysts for the formation of specific glycosidic linkages [14,15]. Other enzymes such as aldolases and sulfotransferases can also be exploited for the synthesis of distinct structures that are critical to oligosaccharide functions.

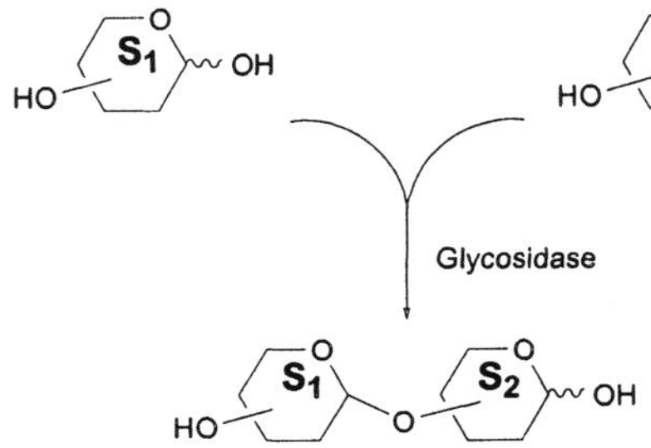

Glycosidase

Scheme 1. Protocol for glycosidase-based synthesis of oligosaccharide

Glycosidases are responsible for glycan processing reactions that take place during glycoprotein synthesis. The physiological function of these enzymes is the cleavage of glycosidic linkages. However, under controlled conditions, glycosidases can be used to synthesize glycosidic bonds (Scheme 1) and have been employed as catalysts in oligosaccharide synthesis. Glycosidases are widely available, robust, and require only inexpensive donor substrates. Although glycosidases are generally stereospecific, they have only weak regiospecificity, which may result in the formation of multiple products.

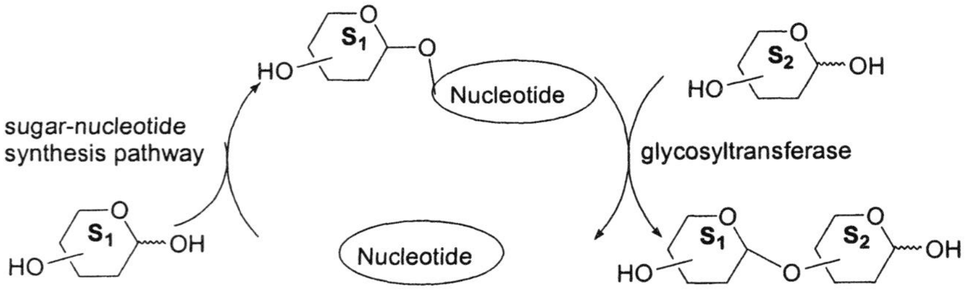

sugar-nucleotide synthesis pathway

Nucleotide

glycosyltransferase

Nucleotide

Scheme 2. Biosynthesis of oligosaccharide by glycosyltransferase

Glycosyltransferases of the Leloir pathway are responsible for the synthesis of most cell-surface glycoforms in mammalian and prokaryotic systems. These enzymes transfer a given carbohydrate from the corresponding sugar nucleotide donor substrate to a specific hydroxyl group of the acceptor sugar (Scheme 2).

Glycosyltransferases exhibit very strict stereospecificity and regiospecificity. Moreover, they can transfer with either retention or inversion of configuration at the anomeric carbon of the sugar residue. A large number of eukaryotic glycosyltransferases have been cloned to date and used in large-scale syntheses of oligosaccharides [14,16,17] with exquisite linkage and substrate specificity. It should be noticed that that such a considerable number of The great majority of mammalian enzymes utilize nine common sugar nucleotides as donor substrates. Glucosyl-, galactosyl, and xylosyltransferases employ substrates activated with uridine diphosphate as the anomeric leaving group (UDP-Glc, UDP-GlcNAc, UDP-GlcA, UDP-Gal, UDP-GalNAc, and UDP-Xyl), whereas fucosyl- and mannosyltransferases utilize guanosine diphosphate (GDP-Fuc and GDP-Man). Sialyltransferases are unique in that the glycosyl donor is activated by cytidine monophosphate (CMP-Neu5Ac).

Recent advances in the area of enzymatic oligosaccharide synthesis are driven by the identification and cloning of a large number of bacterial glycosyltransferases with many different donor, acceptor and linkage specificities [18-22]. Most eukaryotic glycosyltransferases are not active within prokaryotic expression systems because of the absence of post-translation modifications including glycosylation. Bacterial glycosyltransferases, on the other hand, are normally not glycosylated proteins. It has been demonstrated that these enzymes are more easily expressed in soluble and active form in prokaryotic expression system such as *E. coli*. In addition, bacterial glycosyltransferases have relatively broader acceptor substrate specificities, offering potentially important advantages over mammalian enzymes in the chemoenzymatic synthesis of oligosaccharides and their analogues [23,24].

The availability of new glycosyltransferases has increased the demand for sugar nucleotides, which have been a problem in the production of oligosaccharides and glycoconjugates. One way to overcome this hurdle is to use multiple enzyme systems with *in situ* sugar nucleotide regeneration from inexpensive starting materials. Since the pioneering work by Wong et al [25] on *in vitro* enzymatic synthesis of *N*-acetyllactosamine with regeneration of UDP-galactose, several glycosylation cycles with regeneration of sugar nucleotides have been developed using either native or recombinant enzymes [26-30].

The isolation of recombinant enzymes is generally a rather laborious operation. A rapidly emerging method for large-scale synthesis of complex carbohydrates is the use of metabolically engineered microorganisms. So far three strategies have been used in developing the several whole-cell biocatalysts.

Bacterial coupling technology. This technology was developed by Kyowa Hakko Kogyo Co. Ltd. in Japan [31-35]. The key in Kyowa Hakko's technology for the large-scale production of oligosaccharides is a *C. ammoniagenes* bacterial strain engineered to efficiently convert inexpensive orotic acid to UTP (Scheme 3). An *E. coli* strain engineered to overexpress UDP-Gal biosynthetic genes including *galK* (galactokinase), *galT* (galactose-1-phosphate uridyltransferase), *galU* (glucose-1-phosphate uridyltransferase), and *ppa* (pyrophosphatase) utilized the UTP in the production of UDP-Gal (72 mM / 21 h). Combining these two strains with another recombinant *E. coli* strain overexpressing the α1,4-galactosyltransferase gene of *Neisseria gonorrhoeae*, created a system efficient at generating high concentrations of globotriose.

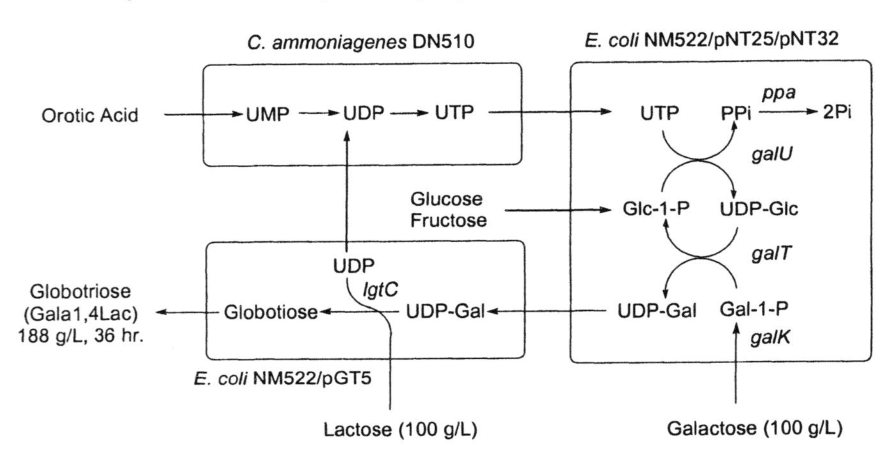

Scheme 3. Large-scale production of oligosaccharides through coupling of engineered bacteria.

The "superbug" technology. This technology was developed by Dr. Peng George Wang's group in the USA. Instead of multiple bacteria strains, the superbug technology uses only one plasmid and a single strain. All the enzymes essential for oligosaccharide synthesis including the glycosyltransferases and the nucleotide sugar regeneration are expressed within one *E. coli* strain. Therefore, this approach avoids unnecessary transport of intermediates in the biosynthetic cycles from one strain to another. Combined with appropriate fermentation process, the technology has been proven to be an efficient method for large-scale production of complex carbohydrates [36,37]. For example, in the synthesis of α-Gal oligosaccharide, the byproduct (UDP) of the galactosylation is recycled to the sugar nucleotide donor (UDP-Gal) through the action of four enzymes namely PykF (pyruvate kinase), GalU (glucose-1-phosphate uridylyltransferase), GalT (galactose-1-phosphate uridylyltransferase) and GalK (galactokinase). All these genes together with the α-1,3-galactosyltransferase gene were incorporated into a single plasmid and transformed into an *E. coli* NM522 strain. Through fermentation, gram-scale production of α-Gal oligosaccharide has been achieved in high yields (50-60%) from simple starting materials such as galactose, lactose and PEP (phosphoenolpyruvate, Scheme 4).

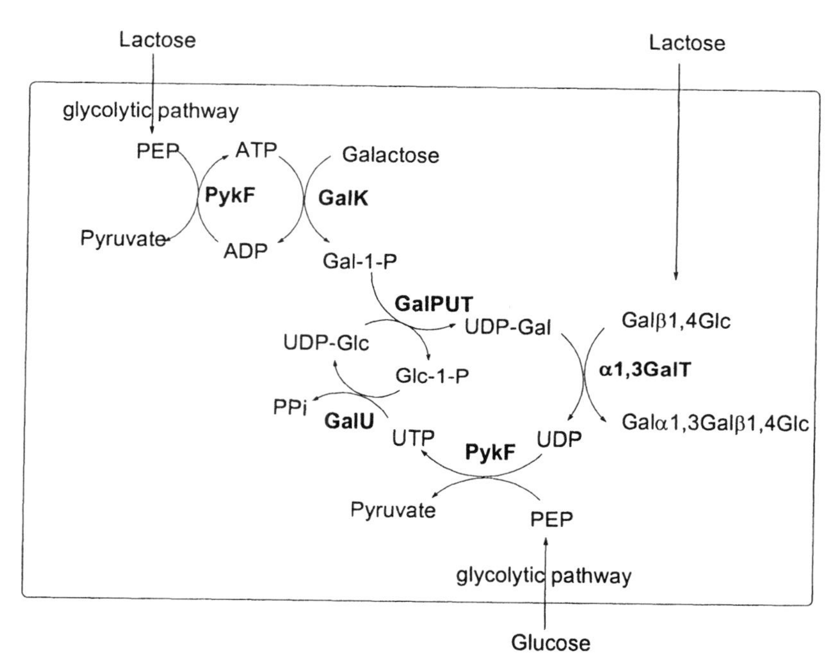

Scheme 4. Wang's superbug technology for synthesis of Galα1,3Lac

The "living factory" technology. This technology was developed by Dr. Samain's group in France. This technology makes use of the bacteria host cell's own ability to produce nucleotide sugars, while the bacteria are simply engineered to incorporate the required glycosyltransferases (Scheme 5). In high-cell-density cultures, the oligosaccharide products accumulate intracellularly and have been shown to reach levels on the gram / liter scale [38-43].

It is becoming clear that no single approach is ideal for the synthesis of carbohydrates. The many approaches currently in use have came from scientists of varying backgrounds including organic chemistry, biochemistry, genetic engineering as well as others. The chapters that follow detail current approaches to the synthesis of carbohydrates. The methods described take advantage of combined chemo-enzymatic syntheses, the availability and ability to manipulate enzymes, genetic engineering of microorganisms as well as automated syntheses that have so successfully been applied to the syntheses of of peptides and oligonucleotides. It is easily recognized that whatever further developments may be necessary, they will require a highly multidisciplinary effort that encourages collaborations between many fields in both academia and industry.

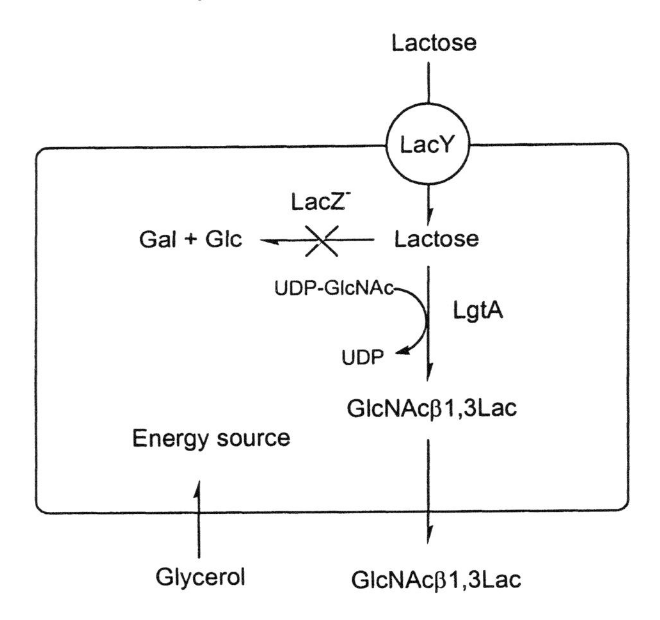

Scheme 5. Production of trisaccharide GlcNAcβ1,3Lac by *E. coli* JM109 expressing the *lgtA* gene that encodes a β-1,3-*N*-acetylglucosaminyltransferase.

Literature Cited:

1. Hundle, B. S., O'Brien, D. A., Alberti, M., Beyer, P., and Hearst, J. E. (1992) *Proc. Natl. Acad. Sci. USA* **89**, 9321-9325
2. Galili, U. (1998) *Sci & Med.* **5**, 28-37
3. Good, A. H., Cooper, D. K., Malcolm, A. J., Ippolito, R. M., Koren, E., Neethling, F. A., Ye, Y., Zuhdi, N., and Lamontagne, L. R. (1992) *Transplant. Proc.* **24**, 559-562
4. Groth, C. G., Korsgren, O., Tibell, A., J., T., Moller, E., Bolinder, J., Ostman, J., Reinholt, F. P., Hellerstrom, C., and Andersson, A. (1994) *Lancet* **344**, 1402-1404
5. Platt, J. L. (1998) *Nature* **392**, 11-17
6. Brenden, R. A., Miller, M. A., and Janda, J. M. (1988) *Rev. Infect. Dis.* **10**, 303-316

7. Rudd, P. M., Endo, T., Colominas, C., Groth, D., Wheeler, S. F., Harvey, D. J., Wormald, M. R., Serban, H., Prusiner, S. B., Kobata, A., and Dwek, R. A. (1999) *Proc. Natl. Acad. Sci. USA* **96**, 13044–13049
8. Jackson, M. P. (1990) *Microb. Pathog.* **8**, 235-242
9. Nataro, J. P., and Kaper, J. B. (1998) *Clin. Microbiol. Rev.* **11**, 142-201
10. Paton, J. C., and Paton, A. W. (1998) *Clin. Microbiol. Rev.* **11**, 450-479
11. Messner, P., and Schaffer, C. (2003) *Fortschr. Chem. Org. Naturst.* **85**, 51-124
12. Flowers, H. M. (1978) *Methods Enzymol* **50**, 93-121
13. Koeller, K. M., and Wong, C. H. (2000) *Chem Rev* **100**, 4465-4494
14. Palcic, M. M. (1999) *Curr Opin Biotechnol* **10**, 616-624
15. Crout, D. H., and Vic, G. (1998) *Curr Opin Chem Biol* **2**, 98-111
16. Wymer, N., and Toone, E. J. (2000) *Curr Opin Chem Biol* **4**, 110-119
17. Ichikawa, Y., Wang, R., and Wong, C. H. (1994) *Methods Enzymol* **247**, 107-127
18. Johnson, K. F. (1999) *Glycoconj J* **16**, 141-146
19. Shao, J., Zhang, J., Kowal, P., Lu, Y., and Wang, P. G. (2002) *Biochem Biophys Res Commun* **295**, 1-8
20. Shao, J., Zhang, J., Kowal, P., and Wang, P. G. (2002) *Appl Environ Microbiol* **68**, 5634-5640
21. Zhang, J., Kowal, P., Fang, J., Andreana, P., and Wang, P. G. (2002) *Carbohydr Res* **337**, 969-976
22. Blixt, O., van Die, I., Norberg, T., and van den Eijnden, D. H. (1999) *Glycobiology* **9**, 1061-1071
23. Sharon, N., and Ofek, I. (2000) *Glycoconj J* **17**, 659-664
24. Sharon, N., and Ofek, I. (2002) *Crit Rev Food Sci Nutr* **42**, 267-272
25. Wong, C. H., Haynie, S. L., and Whitesides, G. M. (1982) *J. Org. Chem.* **47**, 5416-5418
26. Fang, J.-W., Li, J., Chen, X., Zhang, Y.-N., Wang, J.-Q., Guo, Z.-M., Brew, K., Wang, P. G. (1998).*J. Am. Chem. Soc.* **120**, 6635-6638.
27. Haynie, S. L., Whitesides, G. M. (1990.) *Appl. Biochem. Biotechnol.* **23**, 155-170
28. Hokke, C. H., Zervosen, A., Elling, L., Joziasse, D. H., van den Eijnden, D. H. (1996.) *Glycoconj. J.*, 687-692
29. Ichikawa, Y., Liu, L. C. J., Shen, G. J., Wong, C. H. (1991.) *J. Am. Chem. Soc.* **113**, 6300-6302
30. Zervosen, A. a. E., L. (1996) *J. Am. Chem. Soc.* **118**, 1836-1840.
31. Koizumi, S., Endo, T., Tabata, K., and Ozaki, A. (1998) *Nat Biotechnol* **16**, 847-850
32. Endo, T., Koizumi, S., Tabata, K., Kakita, S., and Ozaki, A. (1999) *Carbohydr Res* **316**, 179-183

33. Endo, T., Koizumi, S., Tabata, K., and Ozaki, A. (2000) *Appl Microbiol Biotechnol* **53**, 257-261
34. Endo, T., and Koizumi, S. (2000) *Curr Opin Struct Biol* **10**, 536-541
35. Endo, T., Koizumi, S., Tabata, K., Kakita, S., and Ozaki, A. (2001) *Carbohydr Res* **330**, 439-443
36. Chen, X., Liu, Z., Zhang, J., Zhang, W., Kowal, P., and Wang, P. G. (2002) *ChemBioChem* **3**, 47-53
37. Chen, X., Zhang, J., Kowal, P., Liu, Z., Andreana, P. R., Lu, Y., and Wang, P. G. (2001) *J Am Chem Soc* **123**, 8866-8867
38. Antoine, T., Priem, B., Heyraud, A., Greffe, L., Gilbert, M., Wakarchuk, W. W., Lam, J. S., and Samain, E. (2003) *Chembiochem* **4**, 406-412
39. Dumon, C., Priem, B., Martin, S. L., Heyraud, A., Bosso, C., and Samain, E. (2001) *Glycoconj J* **18**, 465-474
40. Priem, B., Gilbert, M., Wakarchuk, W. W., Heyraud, A., and Samain, E. (2002) *Glycobiology* **12**, 235-240
41. Bettler, E., Samain, E., Chazalet, V., Bosso, C., Heyraud, A., Joziasse, D. H., Wakarchuk, W. W., Imberty, A., and Geremia, A. R. (1999) *Glycoconj J* **16**, 205-212
42. Samain, E., Chazalet, V., and Geremia, R. A. (1999) *J Biotechnol* **72**, 33-47
43. Samain, E., Drouillard, S., Heyraud, A., Driguez, H., and Geremia, R. A. (1997) *Carbohydr Res* **302**, 35-42

Chapter 2

Exploring Carbohydrate Diversity through Automated Synthesis

Obadiah J. Plante

Ancora Pharmaceuticals, 100 Cummings Center, Suite 419E, Beverly, MA 01915

Automated synthesis technologies have become valuable tools in drug discovery. The application of automated synthesis methods to the assembly of oligonucleotides and oligopeptides was critical for the development of the genomic and proteomic fields. Historically, complex carbohydrates have been extremely challenging to synthesize owing to their structural complexity. Recently, novel synthesis strategies have expanded the use of automated solid phase methods to the complex carbohydrate molecule class. These automated methods have the potential to eliminate the critical bottleneck in carbohydrate synthesis: the assembly of complex carbohydrates.

Introduction

The structural diversity of complex carbohydrates, in the form of glycoconjugates, has been appreciated for many years (*1*). The vast majority of glycoproteins are functionalized with multiple complex carbohydrates, each with the potential to diverge in sequence and thus account for the microheterogeneity that is difficult to interpret in terms of a biological signal (*2*). Carbohydrates are often characterized by highly branched motifs (*3*). Each monosaccharide unit in a complex carbohydrate has multiple sites of attachment to the next sugar unit. Additionally, each glycosidic linkage connecting two sugar units can take on one of two possible isomeric forms. There are over one thousand different trisaccharide combinations possible when the nine mammalian monosaccharides are combined. Although not all of the potential sequences are found in nature, carbohydrates represent the most functionally dense class of biopolymers.

Complex carbohydrates also display variation in the patterns of functional groups they are modified with. Glycosaminoglycans, for instance, consist of long linear polysaccharide chains functionalized with sulfate groups at specific positions (*4*). Similarly, members of the Lewis family of carbohydrates are often sulfated in one or more positions within each complex carbohydrate (*5*). Glycosylphosphatidyl inositol molecules (GPIs), on the other hand, contain phosphoethanolamine groups as hydroxyl modifications and an additional source of diversity (*6*). Due to the structural complexity of both carbohydrate sequences and post-synthetic modifications, researchers have devoted years of effort towards the development of new and improved synthetic methods. The richness in sequence diversity has precluded their widespread use in drug development due to the inability to produce a wide variety of complex carbohydrates (*7*). The application of solution-phase methods, enzymatic methods and, more recently, solid-phase methods have allowed for the synthesis of numerous complex carbohydrates.

Current practices in carbohydrate production

There are several methods commonly used throughout the carbohydrate field for the production of complex carbohydrates. The three most utilized methods for accessing carbohydrates for biological studies are: purification from natural sources, enzymatic synthesis, and chemical synthesis. Purification from natural sources is extremely difficult and time consuming due to the microheterogeniety of glycoconjugates (*8*). This results in the isolation of small quantities of material generated and at high cost (*9*). This is a tedious process that requires significant expertise in analytical chemistry and biochemistry in addition to access to sophisticated instrumentation. Currently, the majority of carbohydrate-based therapeutics on the market today is manufactured using biologically derived material due to the lack of other available methods.

Enzymatic synthesis relies on the high specificity of glycosyltransferase-mediated glycosylation reactions (*10*). The high substrate specificity of these enzymes allows the reactions to proceed within complex mixtures and without the need to use protecting groups. While this method has successfully produced a number of structures, only certain carbohydrate linkages can be prepared due to limited substrate specificity of these enzymes. Due to the inherent enzyme specificity, enzymatic methods have been most commonly used for the production of specific carbohydrates (*11*). Enzymatic carbohydrate synthesis has seen the most success by researchers looking to modify the carbohydrate domain of glycoproteins, rather than by those investigating pure carbohydrates.

Traditional carbohydrate synthesis relies on the coupling of a fully protected donor sugar to an acceptor sugar bearing a free hydroxyl group (*12*). Following purification, a unique protecting group is removed to yield a new acceptor sugar. This iterative process of coupling, purification, deprotection, and re-purification performs elongation steps. Because purification is required after each step, synthesis of carbohydrates is extremely time consuming and costly, requiring an average one-two weeks of work for each monosaccharide unit in a complex carbohydrate chain. For example, the synthesis of a pentasaccharide requires nine distinct purification steps (after 5 coupling steps and 4 deprotection steps) and overall yields are routinely in the 0.1-1% range. Despite the tedious nature of the process, solution-phase chemical synthesis is by far the method of choice for carbohydrate synthesis due to the flexible nature of the design (*13*). Any variety of terminal modifications can be performed using a solution-phase synthesis strategy and there is a tremendous amount of literature precedent for the generation of the necessary carbohydrate building blocks (*14*).

Automated solid-phase carbohydrate synthesis

Automated solid phase synthesis, which has enabled synthesis libraries of peptides (*15*) and nucleic acids (*16*) and their use in drug discovery, is emerging as a useful tool for carbohydrate production (*17*). This technology has been difficult to develop largely due to the aforementioned inherent complexity of carbohydrates (*18*). Both nucleic acids and proteins can be assembled only in head-to-tail chains, whereas a chain of carbohydrates has the potential to elongate at any one of five different points on the sugar and these linkages can adopt one of two isomers. The higher degree of complexity in carbohydrate sequences necessitates a sophisticated protecting group strategy and reliable coupling methods (*19*). Incorporating these attributes into an automated synthesis setting where multiple reaction conditions can be screened simultaneously will enhance the further development of current methods. The realization of parallel, automated solid-phase carbohydrate synthesis would enable the production of diverse carbohydrate libraries for biological testing. The initial efforts towards the development of automated solid-phase synthesis methods are described below.

Synthesis strategy

There are two options for the elongation of carbohydrates using a solid-phase strategy: the donor-bound or acceptor-bound approach. In the donor bound method, a monosaccharide donor is immobilized to a solid-support (*20*). Elongation is accomplished via activation of the donor in the presence of an excess of acceptor. The unreacted acceptor and activation side-products are washed away and repetitive washing steps purify the support-bound carbohydrate. The donor-bound method has been used for the synthesis of several saccharides, however, the tendency for highly reactive donors to undergo unproductive reactions leads to a buildup of support-bound side-products. This has limited the utility of the donor-bound method for the assembly of polysaccharides. The alternative method, the acceptor-bound approach, follows the same rationale as the original solid-phase peptide process that is still used today (Figure 1) (*15*). The reactive building block (donor) is delivered in solution to a solid-support bearing a nucleophilic group (acceptor). Reaction with the support provides a support-bound carbohydrate and the side-products and reagents are removed by simple washing manipulations. Elongation is performed by differentiation of a unique position within the carbohydrate and coupling to another building block (*21*). The acceptor-bound strategy allows for the use of excess donor to drive reactions to completion while at the same time not building up any donor-related side-products on the solid-support (*22*).

Automation

The tedious process of controlling reagent addition and reaction conditions of the solid-support reactions as well as washing of the resin has been obviated by a number or automated synthesizers for the synthesis of peptides, nucleic acids and small molecules. The iterative process of elongation/differentiation, albeit in our case with carbohydrates, is an ideal strategy for the use of automated synthesis methods (*23*). We applied an acceptor-bound strategy to the synthesis of polysaccharides utilizing a modified peptide synthesizer. An ABI-433A was reconfigured for carbohydrate synthesis and new coupling cycles were developed.

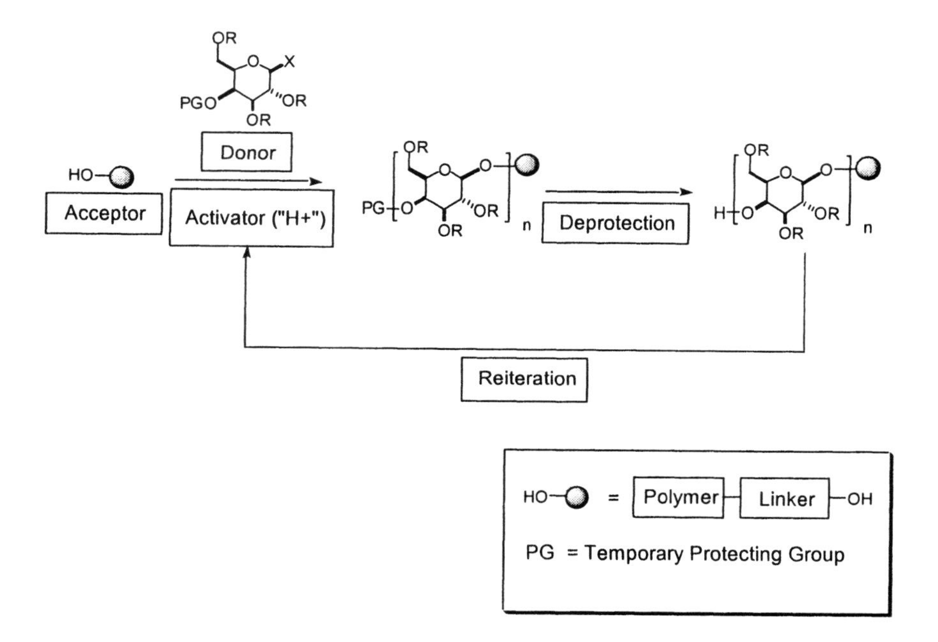

Figure 1. Acceptor-bound carbohydrate synthesis

Glycosyl trichloroacetimidates

The model system for evaluation of the automated solid-phase carbohydrate synthesis process was the assembly of poly-α(1→2)mannosides (Figure 2) (*24*). Glycosyl trichloroacetimidate **2** was used bearing a removable acetate protecting group at C2 (*25*). Glycosylation was performed on a polystyrene support that was functionalized with an octenediol linker (*21*). The extent of either elongation or acetate protecting group removal could be followed by removal of a sample of resin from the reaction mixture and cleavage (olefin metathesis + ethylene) to afford *n*-pentenyl glycosides. Authentic standards of 2-*O*-acetate and 2-OH mono-pentasaccharides were prepared via solution-phase methods to provide reference materials for HPLC analysis of the reaction mixture (*26*). A coupling cycle was developed to perform: 1. Resin washing, 2. Donor and Activator delivery, 3. Vortexing, 4. Resin washing, 5. Acetate deprotection. At the end of a synthesis, the carbohydrate is chemically cleaved from the solid-support and analyzed by HPLC.

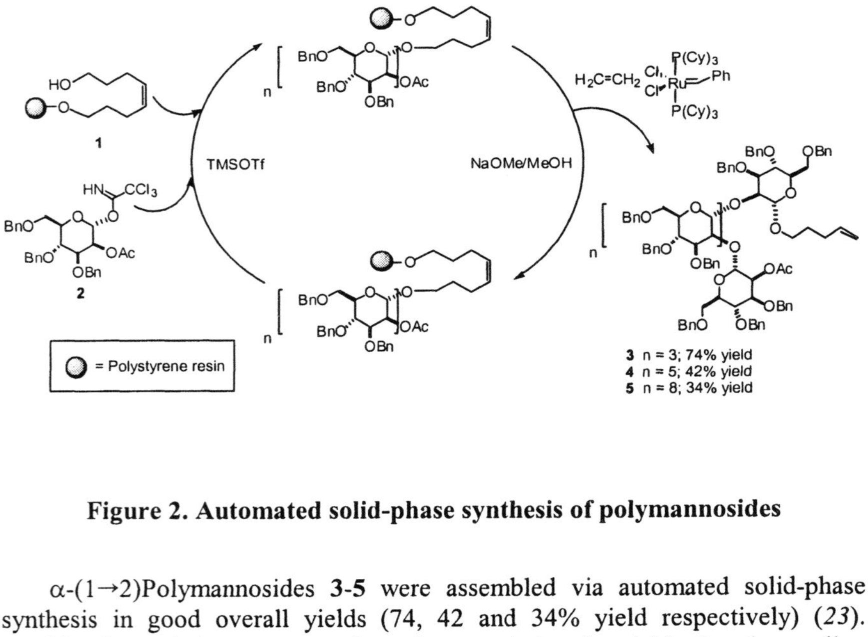

Figure 2. Automated solid-phase synthesis of polymannosides

α-(1→2)Polymannosides **3-5** were assembled via automated solid-phase synthesis in good overall yields (74, 42 and 34% yield respectively) (*23*). Double glycosylations were performed to maximize the yield of each coupling event. Another feature of the coupling cycle that was essential for optimal conversion at the deprotection step was the use of a mixed solvent system (9:1, CH_2Cl_2:MeOH) to ensure proper swelling of the polystyrene resin. The optimized coupling cycle and reaction conditions allowed for average per step efficiency of >95%. The highly efficient process was verified by comparison of a 2D-NMR of pentamannoside **3** (independently prepared) with 2D high resolution magic angle spinning NMR of a resin bound pentasaccharide. The nearly identical spectra confirmed the overall efficiency of the carbohydrate assembly process (*23*). Notably, the synthesis of penta-decamannosides was routinely accomplished with one day. This represents a major acceleration in the rate of carbohydrate synthesis when compared to solution-phase chemical methods that often require weeks to months of diligent effort.

Glycosyl phosphates

In addition to the use of glycosyl trichloroacetimidate donors, a coupling cycle for the use of glycosyl phosphate donors was developed. Glycosyl phosphates are highly reactive building blocks that are often most effective at lower temperatures (*27,28,29*). To accommodate low temperature glycosylations a double-walled reaction vessel was designed (*30*). An internal reaction chamber that was fitted to the synthesizer was connected to an external temperature control device.

Glycosyl phosphate building blocks **6** and **7**, bearing levulinic esters as temporary protecting groups on C6, were utilized in the synthesis of dodecasaccharide **8** (*23*). Alternating glycosylations performed at –15°C with hydrazine deprotections performed at +15°C proved to be a reliable and efficient coupling cycle, affording dodecasaccharide **8** in >50% yield (94% per step). The high stepwise yields and rapid reaction times (dodecasaccharide **8** was synthesized in one day) make glycosyl phosphates ideal building blocks for automated solid-phase carbohydrate synthesis.

Integrated donor method

There are a wide variety of possible glycosyl donor strategies to use when assembling complex carbohydrates (*31*). To date, glycosyl trichloroacetimidates and glycosyl phosphates have been used exclusively for automated solid-phase synthesis owing to their high reactivity and their activation under homogenous reaction conditions. We developed a synthesis strategy for the assembly of **11** that integrates the use of each class of glycosyl donor. The coupling cycles developed for the assembly of both glycosyl trichloroacetimidates and glycosyl phosphates were integrated for the synthesis of trisaccharide **11** (Figure 4) (*23*). Repetitive assembly and deprotection of building blocks **2**, **9** and **10** were effective for the construction of trisaccharide **11**. The integrated coupling cycle strategy offers a flexible method for the assembly of complex carbohydrates using two different types of glycosyl donors and temporary protecting groups.

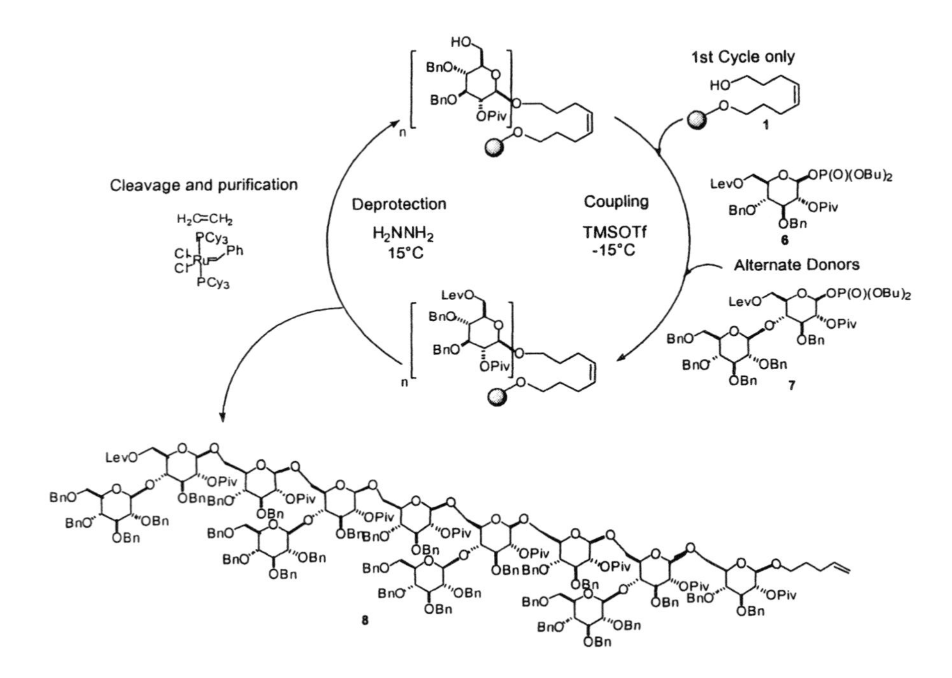

Figure 3. Automated solid-phase synthesis with glycosyl phosphates

This represents the first step towards the development of an integrated technology platform capable of interchanging building blocks for the construction of complex carbohydrates.

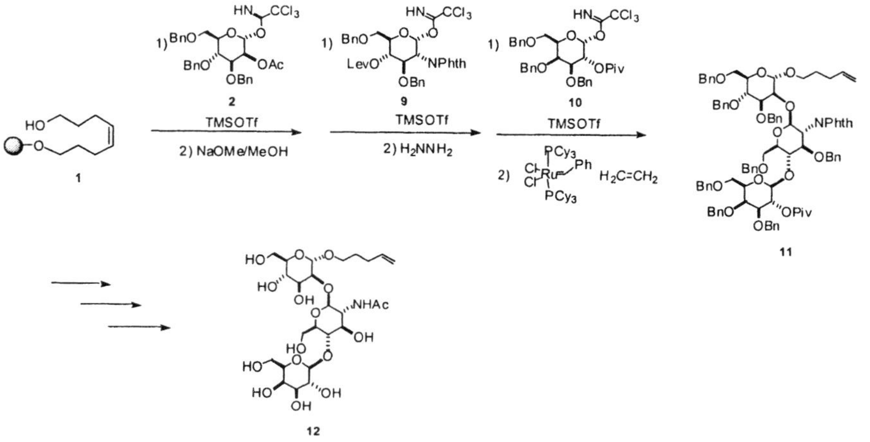

Figure 4. Automated solid-phase synthesis using glycosyl trichloroacetimidates and glycosyl phosphates.

Tagging and purification of compounds

One of the major limitations to the solid-phase synthesis of biopolymers is the purification of the final product. Throughout the iterative process of elongation and deprotection there are many opportunities for impurities and side-products to form on the solid-support. The most significant purification challenge, at least for carbohydrates, is the presence of deletion sequences in the final mixture cleaved from the solid-support. Deletion sequences arise from either the incomplete glycosylation or deprotection. To overcome this issue we developed a Capping and Tagging system whereby, following a glycosylation event, any unreacted, support-bound acceptor would be functionalized with a Cap containing at functional group that would serve as a handle for removal of any Cap-Tagged sequences (*32*).

An example of the Cap-Tag method is shown in Figure 5. The synthesis of a trisaccharide was performed using single glycosylations both with and without the use of the α-azidobutyric acid tag (A-Tag) following each glycosylation event. The crude isolates from the synthesis using the A-Tag reagent were subjected to phosphine reduction and the resulting amino-compounds were removed via filtration through silica gel or reaction with an isocyanate scavenger resin. Comparison of the HPLC trace of the synthesis run with and without the A-Tag step showed a significant effect. By incorporating the A-Tag into the process, all of the mono- and disaccharide deletion sequences were removed. This has important repercussions for the final purification of compounds synthesized on solid support since the A-Tag method dramatically simplifies the final purification event. An alternative to the A-Tag method, the F-Tag (based on a polyfluorinated silyl protecting group) has also proven useful in the synthesis of branched carbohydrates (33,32).

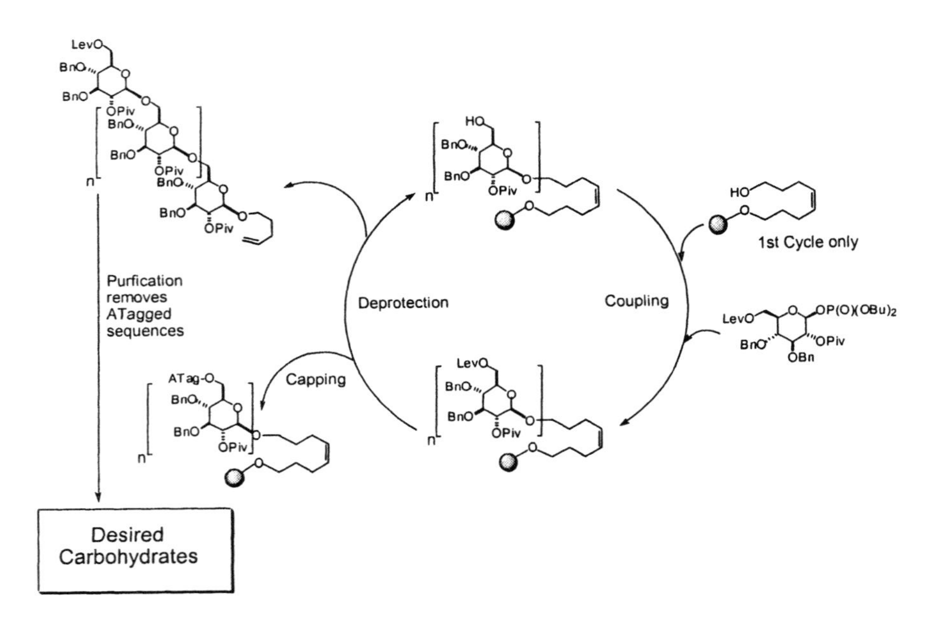

Figure 5. Cap-Tag procedure for purification of biopolymers.

Future directions of automated solid-phase synthesis methods: Application to drug discovery and development

The automated solid-phase synthesis of complex carbohydrates has emerged as a useful tool for the assembly of complex carbohydrates. A limited set of building blocks is required for the synthesis of complex biopolymers in a fully automated fashion. Recent developments include the synthesis of a potential malaria vaccine utilizing automated methods (*34*). In order for this technology to see utility in the drug discovery setting several challenges remain to be overcome. First, the development of a parallel synthesis platform would enable two important applications. It would allow the researcher to optimize the coupling and deprotection conditions for each new building block and it would allow for the assembly of multiple different polysaccharides simultaneously. Secondly, a process chemistry version of a carbohydrate synthesizer would be very attractive for the assembly of lead compounds under GMP conditions. These challenges will take significant effort and creativity to overcome the technical hurdles inherent to each method. The further development of automated solid-phase synthesis technology has direct applications to industry and would form the foundation of a robust platform for the investigation of carbohydrate-based therapeutics.

---

1. Gagneux, P.; Varki, A. *Glycobiology* **1999**, *9*, 747.
2. Orntoft, T.F.; Vestergaard, E.M. *Electrophoresis* **1999**, *20*, 362.
3. Rademacher, T.W.; Parekh, R.B.; Dwek, R.A. *Annu. Rev. Biochem.* **1988**, *57*, 785.
4. Gallagher, J.T.; Lyon, M. in *Proteoglycans: Structure, biology and molecular interactions.* Iozzo, R.V., Ed.; Marcel Dekker, New York, pp 2-61.
5. Komba, S.; Ishida, H.; Kiso, M.; Hasegawa, A. *Bioorg. Med. Chem. Lett.* **1996**, *4*,1833.
6. McConville, M.J.; Ferguson, M.A.J. *Biochem. J.* **1993**, *294*, 305.
7. Dove A: The bittersweet promise of glycobiology. *Nature Biotechnology* **2001**, *19*, 13.
8. Ferro, V.; Fewings, K.; Palermo, M.C.; Li, C. *Carbohydr. Res.* **2001**, *332*, 183.
9. Karamanos, N.K.; Lamari, F. *Biomed. Chromatogr.* **1999**, *13*, 501.
10. Koeller, K.; Wong, C.-H. *Chem. Rev.* **2000**, *100*, 4465.
11. Shao, J.; Zhang, J.; Kowal, P.; Wang, P. G. *Appl. Environ. Microbiol.* **2002**, *68*, 5634.
12. Schmidt, R.R.; Kinzy, W. *Adv. Carbohydr. Chem. Biochem.* **1994**, *50*, 21.

13. Nicolaou, K.C.; Mitchell, H.J. *Angew Chem Int Ed* **2001**, *40*, 1576.
14. Toshima, K.; Tatsuta, K. *Chem Rev* **1993** *93*,1503.
15. Merrifield, R.B. *J. Am. Chem. Soc.* **1963**, *8*, 2149.
16. Caruthers, M.H. *Science.* **1985**, *230*, 281.
17. Plante, O.J.; Palmacci, E.R.; Seeberger, P.H. *Methods in Enzymology* **2003**, in press.
18. Sears, P.; Wong, C.-H. *Science* **2001**, *291*, 2344.
19. For a perspective on other automated carbohydrate synthesis methods see: Tanaka, H.; Amaya, T.; Takahashi, T. *Tetrahedron Lett.* **2003**, *44*, 3053. Zhang, Z.; Ollmann, I.R.; Ye, X.-S.; Wischnat, R.; Baasov, T.; Wong, C.-H. *J. Am. Chem. Soc.* **1999**, *121*, 734.
20. Danishefsky, S.J.; McClure, K.F.; Randolph, J.T.; Ruggeri, R.B. *Science* **1993**, *260*,1307.
21. Andrade, R.B.; Plante, O.J.; Melean, L.G.; Seeberger, P.H. *Org Lett* **1999**, *1*, 1811.
22. Palmacci, E.R.; Plante, O.J.; Seeberger, P.H. *Eur. J. Org. Chem.* **2002**, 595.
23. Plante, O.J.; Palmacci, E.R.; Seeberger, P.H. *Science* **2001**, *291*, 1523.
24. Rademann, J.; Schmidt, R.R. *J Org Chem* **1997**, *62*, 3650.
25. Yamakazi, F.; Sato, S.; Kukuda, T.; Ito,Y.; Ogawa, T. *Carbohydr. Res.* **1990**, *201*, 31.
26. Unpublished results
27. Hashimoto, S.; Sakamoto, H.; Honda, T.; Abe, H.; Nakamura, S.; Ikegami, S. *Tetrahedron Lett.* **1997**, *38*, 8969.
28. Plante, O.J.; Andrade, R.B.; Seeberger, P.H. *Org. Lett.* **1999**, *1*, 211.
29. Plante, O.J.; Palmacci, E.R.; Andrade, R.B.; Seeberger, P.H. *J. Am. Chem. Soc.* **2001**, *123*, 9545.
30. James Glass Inc. of Hanover, MA constructed a specially designed glass reaction vessel.
31. Seeberger, P.H.; Haase, W.-C. *Chem. Rev.* **2000**, *100*, 4349.
32. Palmacci, E.R.; Hewitt, M.C.; Seeberger, P.H. *Angew. Chem. Int. Ed.* **2001**, *40*, 4433.
33. Hewitt, M.C.; Snyder, D.A.; Seeberger, P.H. *J. Am. Chem. Soc.* **2002**, *124*, 13434.
34. Schofield, L.; Hewitt, M.C.; Evans, K.; Siomos, M.-A.; Seeberger, P.H. *Nature* **2002**, *418*, 785.

Chapter 3

Glycosynthesis for Drug Discovery: Globo H Synthesis by OPopS

Shirley A. Wacowich-Sgarbi, David Rabuka, Paulo W. M. Sgarbi, and Yoshi Ichikawa

Optimer Pharmaceuticals, Inc., 10110 Sorrento Valley Road, Suite C, San Diego, CA 92122

Although many chemical approaches have been developed, synthesis of biologically active oligosaccharides and glycoconjugates is still cumbersome because it requires labor-intensive preparation of properly protected sugar derivatives as building blocks. Therefore, the actual work of glycosynthesis in drug discovery is very limited even though many realized that it may have the great potential to generate novel class of drugs. OPopS™ has developed for medicinal chemists to carry out glycosynthesis as efficiently as possible by providing preassembled and well-characterized sugar building blocks. In this chapter, we will describe how OPopS™ can synthesize a complex Globo H hexasaccharide.

Introduction

Carbohydrates are important elements of biomolecules, and display an enormous structural and functional diversity, fulfilling roles as energy sources, structural materials, modulators of protein structure and function, and cellular recognition elements.[1] Carbohydrates are also key components in many drugs, including heparin,[2] anti-cancer agents,[3] and antibiotics such as macrolides[4] and aminoglycosides.[5] It seems natural therefore to include sugar portion in any medicinal chemistry scheme for drug development. However, advances in medicinal chemistry in this field have been limited due to the lack of a robust technology to support the task, with carbohydrate chemistry being perceived as very laborious and time consuming.

Synthesis of a trisaccharide by traditional methods, for instance, normally requires a chemist to start with the laborious preparation of three fully protected sugar derivatives (**II**, **V**, and **IX**), which serve as building blocks, from the corresponding unprotected commercially available sugars (**I**, **IV**, and **VIII**, respectively). Repetitive synthetic steps (selective deprotection of one of the protecting groups in order to unmask the desired OH group and glycosylation: **II** → **III** → **VI** and **VI** → **VII** → **X**) are also time-consuming. The assembly of such a trisaccharide may take months to accomplish. However, if each of the sugar building blocks are available and the assembly can be performed in one-pot, the trisaccharide synthesis can be done in hours.

OPopS™ Technology

OPopS™ (Optimer Programmed One-Pot Synthesis) is a key technology at Optimer Pharmaceuticals Inc, San Dieg, CA. It is a one-pot synthetic procedure for rapid synthesis of oligosaccharides.[6] Because it enables the efficient generation of oligosaccharides with high diversity, it is also an effective synthetic tool for structure activity relationships (SAR) on the sugar portion of drugs. The key elements to the OPopS™ technology are:

1) A wide variety of sugar building blocks (BBs), which presently amount to more than 460 sugar derivatives include:
    - (a) Common sugars which are major components of cell-surface glycoconjugates,
    - (b) Unique sugars which are found in drugs (such as vancosamine and desosamine);
    - (c) Designer sugars, which have been designed/synthesized by our scientists in an attempt to improve the physical and chemical properties of the conjugated drugs.
2) The OptiMer Software, which was designed to select the optimum reaction sequence and BBs required to prepare desired target carbohydrates;[6] and
3) Relative Reactivity Values (RRVs)[6], which are the experimentally determined values relating the reactivities of the BB's to the starndard sugar

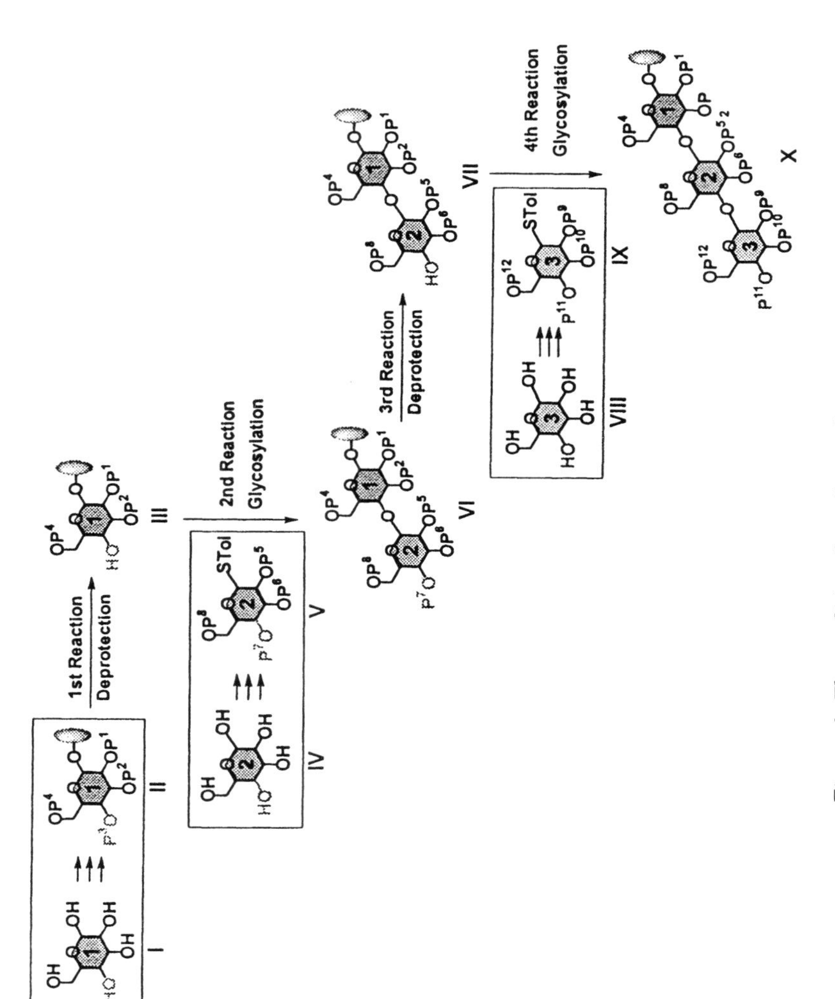

Figure 1. The traditional synthesis of a trisaccharide.

donor. These values are necessary in order to perform a successful one-pot glycosylation and can be manipulated by the choice of protecting groups.

Oligosaccharide synthesis by OPopS™ starts with a careful selection of BBs, aided by OptiMer software, based on the target sequence. The RRV's of each BB are differenciated by at least a factor of 5 for a successful sequential multiple glycosylation operation in one-pot.

The first reaction takes place between the most reactive BB **XI** (highest RRV) and a less reactive BB **XII** (lower RRV) and produces a disaccharide **XIII**. The disaccharide has a thiotolyl group at the anomeric position and thereby can be activated. We have found that the RRV of the formed oligosaccharide, in general, is similar to that of the less reactive BB (in this case **XII**), and therefore the in situ formed disaccharide **XIII** would not be activated under the first glycosylation conditions. After 0.5~1 hr, the third BB **XIV** and additional activator are added to the reaction mixture to activate the disaccharide **XIII** for the second glycosylation. The overall assembly involving 2 glycosylation is usually completed within 2~3 hrs and does not require any deprotection steps. In the following section, we describe how a complex oligosaccahride can be prepared by OPopS™.

Novel therapeutics for cancer: Carbohydrate-based cancer vaccine

Cells express unique carbohydrate structures upon carcinogenic transformation.[7] Therefore, carbohydrate-based antigens offer the potential for a targeted immuno-therapeutic approach to the treatment of certain forms of cancer. Although carbohydrate cancer vaccines have been shown to be safe and to mediate complement-dependent and/or antibody-dependent cell-mediated cytotoxicity, only a limited number of carbohydrate structures have been evaluated in preclinical and clinical settings.[8]

Because these carbohydrates are extremely difficult to synthesize by conventional methodology, only small quantities of these unique oligosaccharides are available for preparing potentially promising antigen constructs in order to maximize the immune response. Biomira Inc. (Edmonton, Canada) was the first to apply a carbohydrate antigen as a cancer vaccine to treat breast cancer patients with a relatively small antigen motif STn to generate antibody.[9] A much larger antigen structure, Globo H hexasaccharide was studied as a cancer vaccine by the Memorial Sloan-Kettering Cancer Center (MSKCC) group led by Danishefsky and Livingston.[10] The success of this approach relies almost entirely on the feasibility of chemical synthesis of oligosaccharide cancer antigen. With OPopS™ technology, Optimer

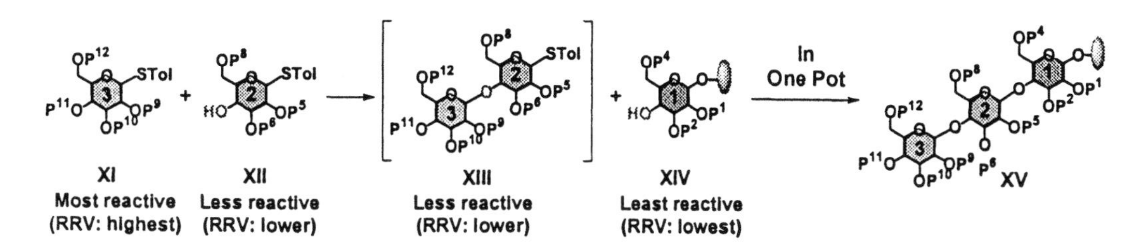

Figure 2. One-pot trisaccharide synthesis using OPopS™.

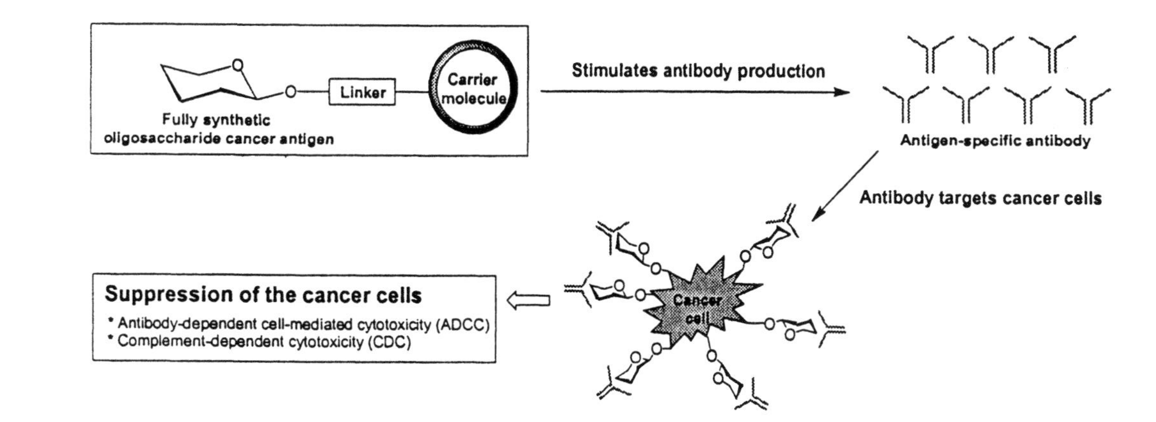

Figure 3. A novel cancer therapy by carbohydrate-based vaccine.

Pharmaceuticals is uniquely positioned to develop these classes of compounds for use as cancer vaccines and has launched a collaborative effort with MSKCC to develop a more effective oligosaccharide-based cancer vaccine.

Globo H Hexasaccharide Synthesis: How quickly Globo H can be prepared

Globo H, a glycosyl ceramide, was isolated and identified as a carbohydrate antigen on prostate and breast cancer cells.[12] The first synthesis of the saccharide moiety of Globo H was reported by Danishefsky *et al.* using a glycal assembly strategy.[13] Lassaletta and Schmidt prepared the hexasaccharide using the trichloroacetimidate methodology.[14] Recently, Zhu and Boons[15]. and the Seeberger group[16] independently reported elegant synthesis of Globo H hexasaccharide. Wong *et al.* reported a first one-pot synthesis of Globo H hexasaccharide.[16]

In all cases, the Globo H hexasaccharide was isolated in a small quantities (<50 mg), therefore, our goal was to establish an improved synthetic scheme that will eventually enable us to generate large quantities for the future clinical development. We also aimed to use OPopS™ to prepare a series of Globo H-related oligosaccharides, the Globo H hexasaccharide **1**, the Globo H terminal tetrasaccharide **6** (Scheme 6) and Gb3 trisaccharide **2**, to be evaluated as an antigen to generate antibody aginst tumor cells.

The Wong's synthesis of Globo H hexasaccharide used three building blocks (Figure 5): the thiofucoside **8**, the trisaccharide **9**, and the lactose building block **10**. and required a rather troublesome α-glycosylation between **9** and a low reactive 4"-OH group of **10**.[16]

One-Pot Approach I involving 3 consequtive glycosylations

We first attempted to limit the building blocks to mono- and disaccharides and experimented the 3 consequtive glycosylations in one-pot to see how the hexasaccharide assembly would be affected. Four building blocks: thiofucoside **8**, disaccharide **11**, thiogalactoside **12**, and the lactose building block **10** (Figure 6). Although the desired β-linkage between the Gal and GalNAc residues was pre-established in the disaccharide building block **11**, the critical α-glycosylation would be examined in this strategy.

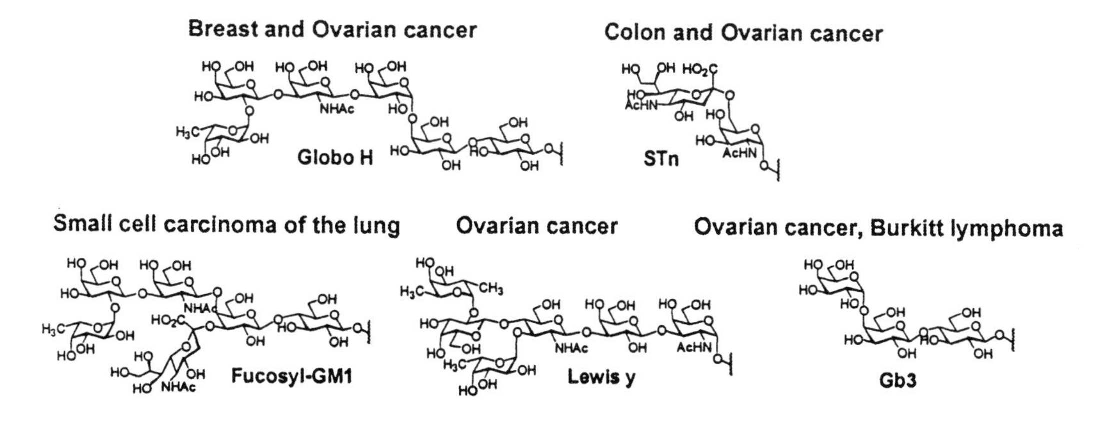

Figure 4. Cancer antigens are specifically expressed on particular tumor cells.

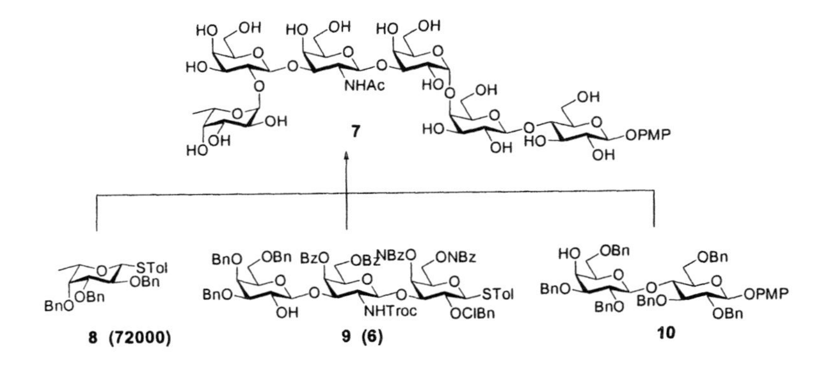

Figure 5. Wong's One-Pot synthesis of Globo H.[16] *Numbers in parentheses are relative reactivity values (RRVs).*

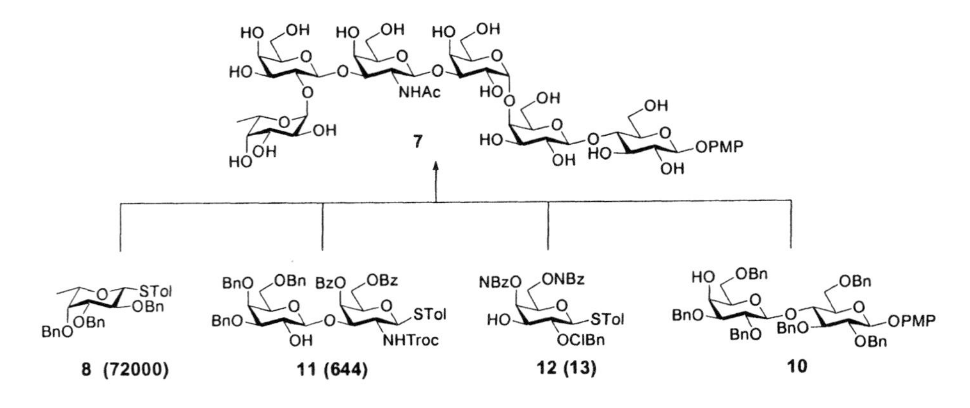

Figure 6. Globo H one-pot synthesis by Approach I. Numbers in parentheses are relative reactivity values (RRVs).

In order to prepare the new building block **11**, thioglycosides donor **13** and acceptor **14** were glycosylated to give disaccharide **15** in 85% yield, which was then treated with hydrazine to deprotect the 2'-O position (Scheme 1). The RRV of this compound relative to peracetyltolylthiomannoside was determined to be 644, which is similar to that of building block **14**, which has a RRV of 850.

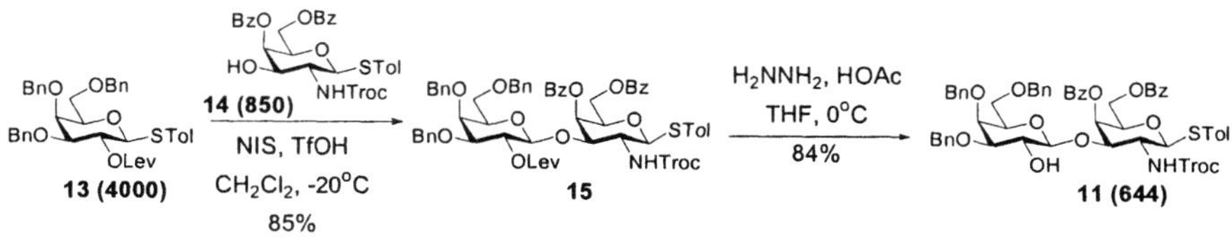

BzO OBz HO STol NHTroc
14 (850)
BnO OBn BnO STol OLev
13 (4000)
NIS, TfOH
CH_2Cl_2, -20°C
85%
BnO OBn BzO OBz BnO O STol OLev NHTroc
15
H_2NNH_2, HOAc
THF, 0°C
84%
BnO OBn BzO OBz BnO O STol OH NHTroc
11 (644)

Scheme 1: *Synthesis of disaccharide building block* ***11****. Numbers in parentheses are relative reactivity values (RRVs).*

The one-pot synthesis of fully protected Globo H derivative **16** is outlined in Scheme 2. In this strategy, the one-pot reaction is used to build up the two α- and one β-linkage during the three sequential glycosylations. The first glycosylation established the Fuc→Gal α-linkage by using donor **8** (2 equiv) and acceptor **11** (1 equiv). The reaction was carried out in dichloromethane at –30°C using N-iodosuccinimide (NIS) and triflic acid as promoters. The next glycosylation established the GalNAc→Gal β-linkage of Globo H by adding the acceptor building block **12** (0.8 equiv), along with AgOTf (0.9 equiv). Instead of using NIS as a promoter, which would inherently produce succinimide as a competing acceptor molecule, AgOTf was added in order to produce the promoter TolSOTf *in situ* from TolSI, which was presumably a side product from the first glycosylation step. Once the second coupling was complete, the final glycosylation step, to make the Gal→Gal α-linkage, was accomplished by adding the final acceptor building block **10** (2 equiv) and AgOTf (1.1 equiv), which generates more promoter *in situ*. After 3 hrs, we were able to isolate the fully protected Globo H hexasaccharide in 15% overally yield (53% yield per coupling). It should be noted that when the one-pot reaction was promoted solely with NIS and TfOH, the overall yield was only 8%. The deprotection of the hexasaccharide **16** followed the procedure reported by Wong *et al.*[16]

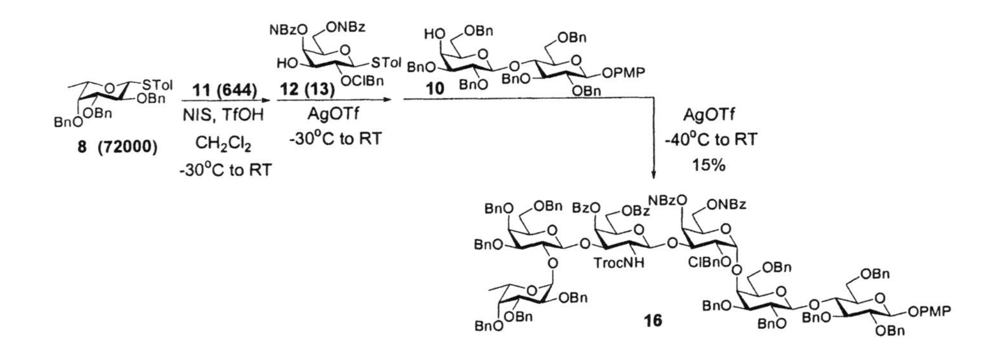

Scheme 2: *Synthesis of Globo H. Numbers in parentheses are relative reactivity values (RRVs).*

We reasoned that the lower overall yield of this approach was very likely due to the fact that an extra coupling step was involved, therefore increasing the chances of undesirable side reactions, and hydrolysis product. Also, the last coupling step involves establishing an α(1→4) linkage to Gal, which is generally a difficult glycosidic linkage to make.

One-Pot Approach II with newly designed BB in which the difficult α-galactosyl linkage built-in

Based on the observation, we designed a new one-pot approach, which would utilize 3 building blocks: mono- (**8**), di- (**11**), and trisaccharide (**18**) in which the difficult α-galactosyl linkage was already incorporated in order to improve the overall yield of oligosaccharide assembly in one-pot, thereby requiring two coupling operations (Figure 5). Trisaccharide **18** is an important building block: not only does it have the pre-established Galα(1→4)Gal linkage, which is difficult to make, but it is also an important structure since it is found in all glycosphingolipids with the globotetraosyl core (GalNAcβ(1→3)Galα(1→4)Galβ(1→4)Glc). Therefore such a building block would be very valuable for the preparation of other representative glycosphingolipid structures. In addition, because the trisaccharide **18** is also known as the Gb3 tumor-associated glycosphingolipid antigen, expressed on Burkitt lymphoma and ovarian cancer cells,[17] it may be useful to be evaluated as a potential cancer vaccine antigen.

The new building block **18** was prepared following the route outlined in Scheme 3. Thioglycoside **20** was glycosylated with the lactose building block **19** and the chloroacetate was selectively removed with thiourea to give building block **18**.

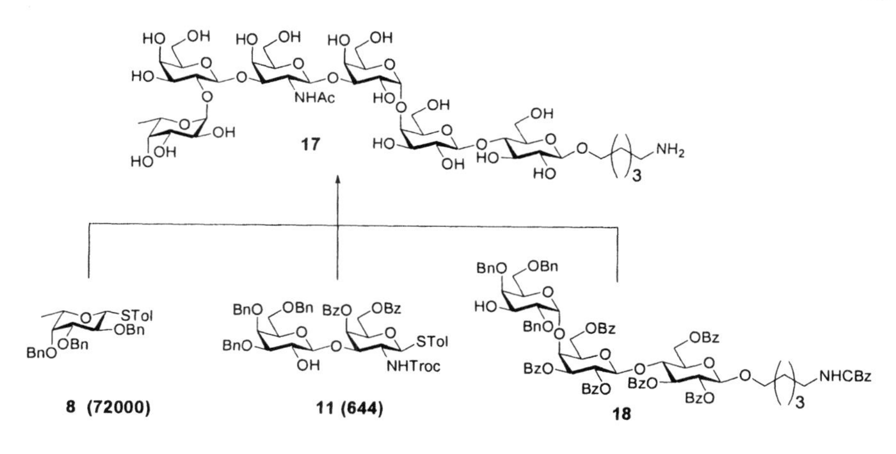

Figure 7. *Globo H one-pot synthesis by Approach II. Numbers in parentheses are relative reactivity values (RRVs).*

1) NIS, TfOH
20
2) Thiourea
19
18

Scheme 3: *Preparation of Gb3 trisaccharide building block **18**.*

The one-pot synthesis of fully protected Globo H derivative **21** is outlined in Scheme 4. In this strategy, the one-pot reaction is used to build up the one α-linkage and one β-linkage, therefore involving two sequential glycosylations with building blocks **8**, **11**, and **18**. The Fuc→Gal α-linkage was first established by reacting donor **8** (1.3 equiv) and acceptor **11** (1.0 equiv) in dichloromethane at –40°C using N-iodosuccinimide (NIS, 1.3 equiv) and triflic acid as promoters. The next step involved establishing the GalNAc→Gal β-linkage by adding the final acceptor **18** (0.8 equiv), along with more NIS (1.3 equiv). A protected Globo H hexasaccharide **21** was efficiently assembled in one-pot within 3 hrs and was isolated in 62%.

The fully protected hexasaccharide derivative **21** was treated with Zn in acetic acid to remove the Troc group to generate a free amino derivative, which was then acetylated to give **22** (Scheme 5). During this deprotection stage, we observed partial O-debenzoylation intermediates, as well as recovered starting material. After O-debenzoylation and the subsequential hydrogenation of **22**, the Globo H antigen **17** with a modifiable amino group at the aglycon linker to be used for the conjugation was obtained.

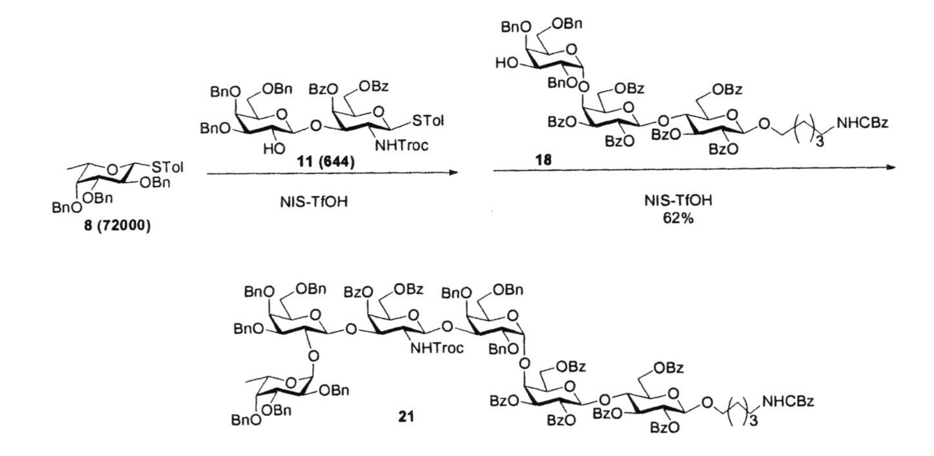

Scheme 4. *Globo H synthesis. Numbers in parentheses are relative reactivity values (RRVs).*

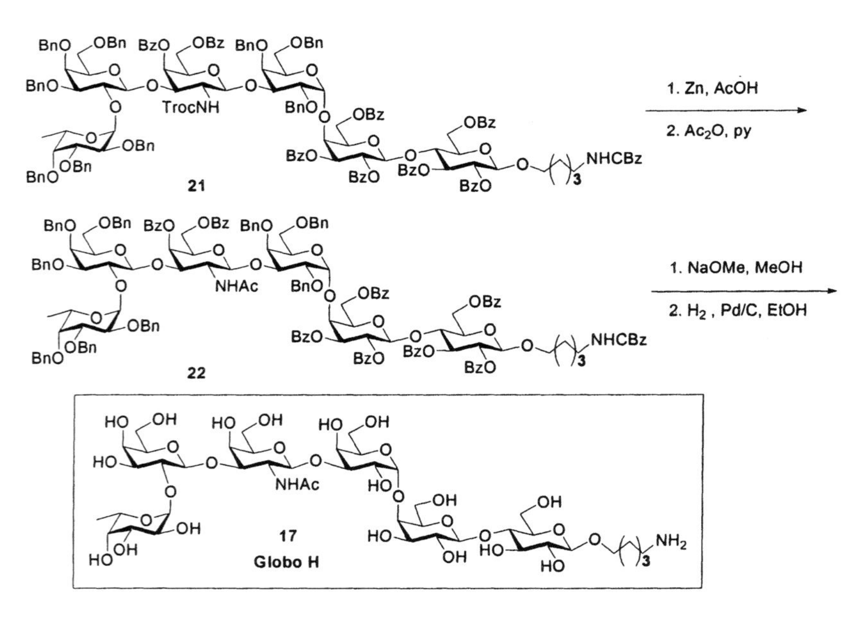

Scheme 5. *Deprotection of the Globo H hexasaccharide.*

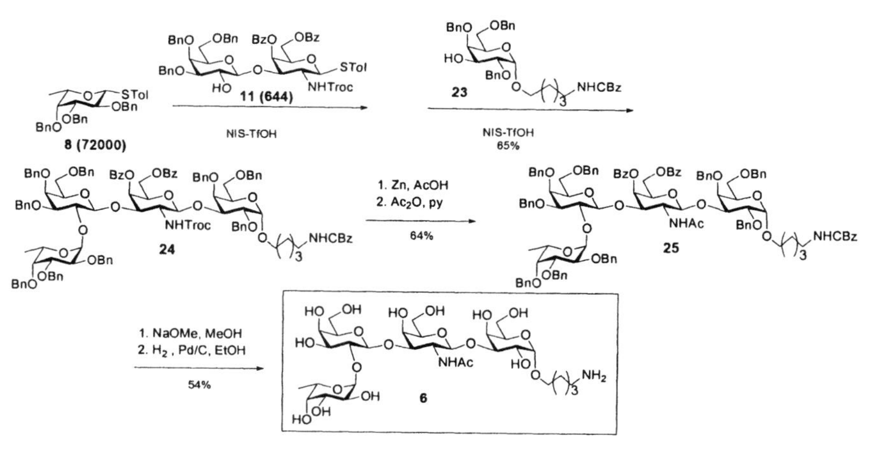

Scheme 6. Synthesis of the Globo H tetrasaccharide. Numbers in parentheses are relative reactivity values (RRVs).

Scheme 7. Deprotection of the Gb3 trisaccharide.

Additional related cancer antigens were also prepared from the sugar building blocks. The terminal tetrasaccharide of Globo H was synthesized in 65% yield by a one-pot reaction involving the sequencial additions of thiofucoside **8**, disaccharide **11** and galactose acceptor **23** (Scheme 6). After converting the NHTroc to an acetamido group and a two step deprotection (Scheme 6), the Globo H tetrasaccharide antigen **6** was obtained. Deacetylation and hydrogenation of the trisaccharide building block **18** gave the Gb3 cancer antigen **26** (Scheme 7).

We have shown efficient synthesis of Globo H-related cancer antigen oligosaccharides by OPopS™. This versatile oligosaccharide synthesis technology will facilitate the exploration of optimizing antigen structure to be used as a novel oligosaccahride-based cancer vaccine and other carbohydrate-based therapeutics. The cancer antigens, Gb3 **26**, Globo H hexasaccharide **17** and tetrasaccharide **6** are being conjugated to an appropriate carrier protein in order to prepare new class of cancer vaccines to be evaluated how effectively a multifaceted immune response would be stimulated.

OPopS™ technology has now provided for the first time an exciting opportunity for chemists to perform carbohydrate-related drug-discovery in antibiotics, anti-inflammatory agents, antithrombotic agents, vaccines, and glycopeptides. OPopS™ technology can also be applied to generate very efficiently oligosaccharide diversity, which would be suitable to establish a glyco array for target identification and diagnosis.

OPopS™for medicinal chemistry

The OPopS™ technology can also be applied to drug discovery, especially novel antibiotics in the macrolides, aminoglycosides, and glycopeptide class of antibiotics, where the carbohydrate moieties are known to play important roles in binding and specificity. Therefore, medicinal chemistry on the sugar components of such classes of drugs has great potential. Our collection of sugar building blocks include sugar derivatives in which key functional groups are moved systematically around the rings that allow sugar-based medicinal chemistry to be performed in a short period time (Figure 4).

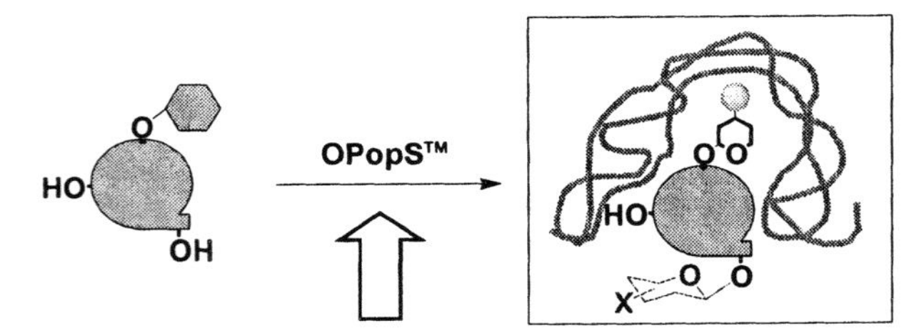

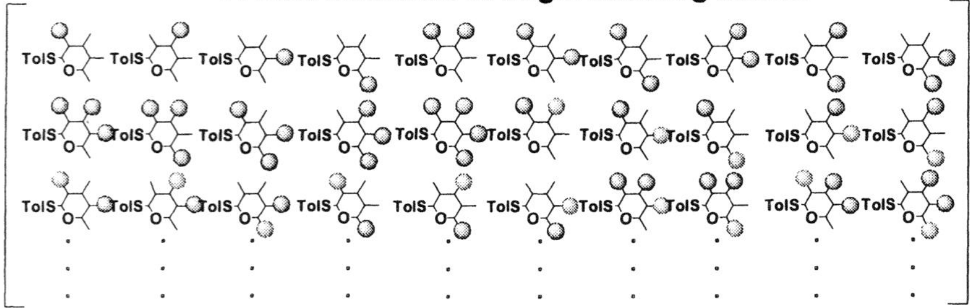

Figure 8. Carbohydrate-based medicinal chemistry by OPopS™

In the case of aminoglycosides, we have applied the OPopS™ technology on a unique aminoglycoside core, prepared from known antibiotic tobramycin, to produce variety of aminoglycoside derivatives. Among them, compound OP-382 was identified as a novel, promising lead that is active against a variety of bacteria strains. Further optimization is now underway in our laboratory.

OPopS™ is also being applied in our macrolide discovery program. We have developed an efficient procedure that allows us to prepare a macrolide aglycon with the 5-OH group available for medicinal chemistry. Glycosylation of the aglycon by the OPopS™ technology yields novel macrolides with a variety of sugar structural diversity, including mono- and disaccharides. We have identified very active macrolides with superior profile against target bacteria. Again, further optimization is now underway in our laboratory.

Presently, we have established the infrastructure of our OPopS™ Technology – we have a wide selection of BBs in our database (>440), with measured RRVs; the Optimer software, which guides the synthesis of oligosaccharides, is being upgraded; and automation of the process is in progress. OPopS™ is a proven technology, demonstrated by the synthesis of several cancer antigens and the successful application to our drug discovery programs. OPopS™ has the great potential as a very efficient technology to perform glycosynthesis for drug discovery.

Acknowledgements: We thank all our colleagues for their work: Chemistry: Alex Romero, Chang-Hsing Liang, Jack Hwang, Jonathan Duffield, Ken Marby, Mayling Cheng, Shirley Leung, Steve Sucheck, Sulan Yao, Youe-Kong Shue, Zhiyuan Zhang; Biology: Farah Babakhani, Jaime Seddon, Mrunali Bairi, Nikki Robert, Pamela Sears, San-Bao Hwang.

References

1. Dwek, R.A. *Biochem. Soc. Trans.* **1995**, *23*, 1-25.

2. Lever, R.; Page, C.P. *Nature Rev. Drug Disc.* **2002**, *1*, 140.

3. Supino, R.; Zunino, F. *Pharmacol. Ther.* **1997**, *76*, 117-124.

4. Omura, S (Ed.) *Macrolide Antibiotics: Chemistry, Biology, and Practice*, 2nd Ed., Academic Press, New York, 2002.

5. Kotra, L.P.; Mobashery, S. *Curr. Org. Chem.*, **2001**, *5,* 193.

6. Zhang, Z.; Ollmann, I. R.; Ye, X.-S.; Wischnat, R.; Baasov, T.; Wong, C.-H. *J. Am. Chem. Soc.* **1999**, *121*, 734.

7. Hakomori, S.; Zhang, Y. *Chemistry & Biology* **1997**, *4*, 97-104.

8. (a) Livingston, P.O.; Zhang, S.; Lloyd, K.O. *Cancer Immunol. Immunother.* **1997**, *45*, 1-9. (b) Livingston, P.O.; Ragupathi, G. *Cancer Immunol. Immunother.* **1997**, *45*, 10-19.

9. (a) MacLean, G.D.; Miles, D.W.; Rubens, R.D.; Reddish, M.A.; Longnecker, B.M. *J. Immunother. Emphasis Tumor Immunol.* **1996**, *19*, 309-319. (b) Holberg, L.A.; Sandmaier, B.M. *Expert Opin. Biol. Ther.* **2001**, *1*, 881-891.

10. (a) G. Ragupathi, T. K. Park, S. Zhang, I. J. Kim, L. Garber, S. Adluri, K. O. Lloyd, S. J. Danishefsky, P. O. Livingston, *Angew. Chem. Int. Ed. Engl.* **1997**, *36*, 125–128. (b) G. Ragupathi, S. F. Slovin, S. Adluri, D. Sames, I. J. Kim, H. M. Kim, M. Spassova, W. G. Bornmann, K. O. Lloyd, H. I. Scher, P. O. Livingston, S. J. Danishefsky, *Angew. Chem. Int. Ed. Engl.* **1999**, *38*, 563–566. (c) S. J. Danishefsky, J. R. *Angew. Chem. Int. Ed.* **2000**, *39*, 836–863. (d) G. Ragupathi, D. M. Coltart, L. J. Williams, F. Koide, E. Kagan, J. Allen, C. Harris, P. W. Glunz, P. O. Livingston, S. J. Danishefsky, *Proc. Natl. Acad. Sci. USA* **2002**, *99*, 13699-13704.

11. (a) R. Kannagi, S. B. Lewery, F. Ishigami, S. Hakomori, L. H. Shevinsky, B. B. Knowles, D. Solter, *J. Biol. Chem.* **1983**, *258*, 8934–8942. (b) E. G. Bremer, S. B. Levery, S. Sonnino, R. Ghidoni, S. Canevari, R. Kannagi, S. Hakomori, *J. Biol. Chem.* **1984**, *259*, 14773–14777.

12. M. T. Bilodeau, T. K. Park, S. Hu, J. T. Randolf, S. J. Danishefsky, P. O. Livingston, S. Zhang, *J. Am. Chem. Soc.* **1995**, *117*, 7840.

13. J. M. Lassaletta, R. R. Schmidt, *Liebigs Ann.* **1996**, 1417-1423.

14. T. Zhu, G.-J. Boons, *Angew. Chem. Int. Ed.* **1999**, *38*, 3495-3497.

15. Bosse, F.; Marcaurelle, L.A.; Seeberger, P.H. *J. Org. Chem.* **2002**, *67*, 6659-6670.

16. Burkhart, F.; Zhang, Z.; Wacowich-Sgarbi, S.; Wong, C.-H. *Angew. Chem. Int. Ed. Engl.* **2001**, *40*, 1274.

17. H. Farkas-Himsley, R. Hill, B. Rosen, S. Arab, C. A. Lingwood, *Proc. Natl. Acad. Sci. USA* **1995**, *92*, 6996-7000.

Chapter 4

Probing the Antigenic Diversity of Sugar Chains

Ruby Wang, Brian J. Trummer, Ellen Gluzman, Chao Deng, and Denong Wang*

Division of Functional Genomics, Columbia Genome Center, College of Physicians and Surgeons, Columbia University, 1150 St. Nicholas Avenue, New York, NY 10032
***Corresponding author: dw8@columbia.edu**

Sugar chains of living organisms are structurally diverse and rich in biological information. The cellular functions of a specific sugar chain depend on its own structural characteristics, as well as on its interaction with other biological molecules that are able to recognize its structure. To facilitate investigations of carbohydrate-mediated molecular recognition, we have established a carbohydrate-based microarray technology. This biochip platform allows a single protein, such as an antibody, a lectin, or other cellular protein, to interact with thousands of carbohydrate molecules of distinct structures in a single assay and thus allows us to study the specificity and cross-reactivity of carbohydrate-protein interactions in a way that was previously impossible. A promising medical application is the detection and characterization of the anti-carbohydrate antibodies that are elicited by a microbial infection or are associated with human diseases with abnormal expression of complex carbohydrates. Such technology is also a powerful means of studying the relationship between structural characteristics of carbohydrates and their antigenic properties. This article discusses the presence of different categories of carbohydrate-based antigenic determinants and the need for microarray technologies to preserve sugar chain conformations on chip.

Carbohydrate-containing macromolecules of biological origin have been recently designated the term "glycomers" (*1, 2*), reflecting the influence of a new scientific discipline, glycomics. Polysaccharides and glycoconjugates of multiple structural configurations, such as glycoproteins, glycolipids, glycosphingolipids, glycosaminoglycans and proteoglycans, are all covered by this term. Cellular glycoconjugates are involved in many biological processes. Documented examples include molecular recognition at fertilization (*3, 4*), cell-cell recognition, adhesion, and cell activation throughout the development and maturation of organisms (*1, 5-7*). Abnormalities in the expression of complex carbohydrates are seen in cancer (*8, 9*), retroviral infections (*10, 11*), and other diseases with a genetic or somatic defect in biosynthesis or metabolism of cellular glycans (*12*). Many carbohydrate molecules serve as the "signatures" of a specific species or strain of microorganism (See Ref (*13, 14*) for reviews). Such carbohydrate structures are important molecular targets for vaccine development and/or diagnosis of infectious diseases (*15-19*).

Glycomics extends proteomics to include carbohydrate-containing biological molecules. This scientific discipline seeks to probe the diversity, decipher the information content, explore the biomedical applications, and examine the biosynthesis of glycomers. Glycomics also strives to facilitate efficient genetic and chemical engineering of these carbohydrate-containing macromolecules and seeks to firmly establish technologies for a large-scale production of such biological molecules. Our current research focuses on developing a biochip-based high throughput technology to explore the antigenic diversity of glycomers and to "decode" the sugar chain-based biological "codes." Additionally, we are developing a technology that can enable the simultaneous detection and immunological characterization of a wide range of microbial infections (*2*).

Rapid progress of the genome-sequencing projects has led to the development of a generation of high throughput technologies for biological and medical research that include nucleic acid- (*20-23*), protein- (*24-26*), and carbohydrate-based microarrays (*2, 27*). These approaches are designed to monitor the expression of genes, proteins, and glycomers on a large scale, and to identify the characteristic patterns of their differential expression in an organism at given conditions. In principle, cDNA microarray and oligonucleotide biochips (*23*) are interchangeable for the purpose of expression analysis, as the detection specificity of these biochips is determined solely by the A::T and C::G base pairing. An on-chip denaturation process is prerequisite in order to allow DNA:DNA or DNA:RNA hybridization and their specific base pairing to take place. This is strikingly different from the protein-based microarrays, in which preservation of protein structure, especially its 3D conformation on chip, is necessary and poses a technical challenge to current protein microarray techgnology. Whether a carbohydrate microarray requires preservation of the conformational properties of carbohydrate molecules on chip is a more complicated issue and will be further explored in the following discussion.

Structural diversity of carbohydrates

While the primary structures of both proteins and carbohydrates are determined by the sequence of their respective monomers, the structure of carbohydrates is further characterized by the variety of their linkages. Unlike proteins that are connected solely by a peptide bond, carbohydrates utilize many possible glycosidic linkages so as to extensively diversify their structures. For example, whereas two amino acids can produce only one possible dipeptide, two molecules of glucose are able to generate eleven different disaccharides. Thus, while nucleic acids and proteins have a single type of covalent bond between their monomers, the structure of a sugar chain is determined by the composition and sequence of its sugar residues, as well as by their glycosidic linkages, and can therefore generate a variety of structures with identical residue compositions and sequences.

Conformational flexibility, or plasticity, is an intrinsic property of sugar chains (*14*). Oligosaccharides and polysaccharides may exist in many different conformations wherein the lower energy forms predominate. Carbohydrates are further diversified by microheterogeneity. Unlike the translation of mRNA messages into proteins, which is precisely conducted by a protein biosynthesis pathway, the synthesis of sugar chains requires multiple enzymes and a complex pathway of biosynthesis. Carbohydrates with microheterogeneity are thus unavoidably generated during their biosynthesis.

Biosignals of carbohydrates

There are a number of biological systems that are able to specifically recognize carbohydrate structures. For example, the enzymatic recognition of dextran and glycogen by proteins of either microbial or mammalian origin is glycosidic linkage-specific. Dextran and glycogen are polymers composed entirely of glucose. Dextrans are produced by microbes, while glycogen is one of the forms of energy storage for mammals. Mammals can degrade α(1,4)-linked glycogen. However, they are unable to break down dextran, which has predominantly α(1,6)-linkages. α(1,6)dextran of relatively low molecular weights can thus be used as a blood expander to maintain normal blood volume and blood pressure in the human body (*28*).

The host's immune system can recognize dextrans and mount specific antibody responses to react with dextrans produced by the microbes (*29-31*). Differing from glycogen, α(1,6)dextrans with molecular weights of over 90,000 are antigenic in human and mouse (*31-33*). Dextran molecules derived from different strains may differ significantly in their glycosidic linkage compositions (*34*). While some dextran preparations are predominantly or solely α(1,6)linked, forming molecules with predominantly linear chain structures; others are composed of multiple glycosidic linkages, including α(1,6), α(1,3), α(1,2), and others, generating heavily branched molecules (*34*). Previous immunological studies have demonstrated that such structural characteristics are detectable by antibodies specific for different antigenic determinants, or epitopes, of dextran molecules (*29, 31*).

Complex carbohydrates serve as carriers of biological information with a capacity larger than that of simple polysaccharides. The specificity of the major blood group types of human red blood cells is determined by the differences in the sugar chains of a glycoprotein or glycolipid on the red blood cell surface. The sugar structures of blood type A and B differ from those of type H by one extra sugar connected to the terminal galactose via an α(1,3)linkage. This extra monosaccharide is acetylgalactosamine in the A blood group substance and galactose in the B group substance. Though differing from each other by only a few atoms, the terminal branches of the sugar chains of blood group substances are the sole determinants of the specificity of blood types and of the barrier of blood transfusions between unmatched donor and recipient.

Many microorganisms can recognize specific sugar chains and use them as receptors that may aid them in entering into certain types of host cells (*35-37*). For example, *Mycoplasma Pneumoniae*, a pathogen of the respiratory tract, attaches to carbohydrate receptors on host erythrocytes or pulmonary fibroblasts in a sialic acid-dependent manner (*35*). This microbe interacts primarily with the sialoglycoprotein, but not asialoglycoprotein, which differs from the former solely by the absence of an $\alpha(2,3)$linked sialic acid. Sugar chips of large capacity and structural diversity are thus useful in pathogen recognition. Identification of such carbohydrate-mediated molecular recognition may lead to a better understanding of the biological relationship between microbes and their mammalian hosts.

Protein glycosylation also occurs in intracellular proteins (*38, 39*). Important nuclear and cytosolic events, such as nuclear transport, cytoskeletal assembly, transcription, and translation, are regulated by protein O-glycosylation at given positions. In fact, the frequency and dynamic nature of O-glycosylation seems to parallel the occurrence of protein phosphorylation. There is often an active and complex interplay between O-GlcNAc and O-phosphate, involving single or reciprocal modification at either identical or adjacent sites. The two post-translational modifications can also occur independently or by influencing each other in a regulated manner. The biological significance of these novel findings has yet to be further investigated.

Antigenic diversity of carbohydrates

Polysaccharides are generally "complete antigens" in that they are able to induce reactive antibodies in certain species. By contrast, oligosaccharides may serve as components or antigenic determinants of an antigen but are unable to evoke an immune response in their free, unconjugated forms. A carbohydrate antigen may express more than one antigenic determinant, including dominant antigenic determinants, as well as minor antigenic determinants. The former is considered as the “key” structure for the induction of an immune response. Identification of such structures is important for the development of carbohydrate-based diagnostic tools or vaccines.

The nature of carbohydrate-based antigenic determinants has been extensively studied (*13, 14*). In the 1950s, the late Professor Elvin Kabat and his colleagues studied human antibodies to α(1,6)dextran and characterized their binding sites. They demonstrated in this system that a carbohydrate epitope may vary from a lower limit of one to two glucoses to an upper limit of six to seven (*32, 40*). Immunochemical mapping of the combining sites of two monoclonal myeloma proteins specific for α(1,6)dextran, W3129 and QUPC52, further established the existence of antibody bining sites, which are specific for either the internal linear chain or the terminal structures of carbohydrate molecules. These carbohydrate structures and their antibody recognition are schematically shown in Figure 1.

W3129 had a site saturated by isomaltopentaose (IM5), whereas QUPC52 accommodated isomaltohexaose (IM7). The terminal non-reducing glucose contributes 50-60 % of the total binding energy with W3129 but less than 5 % with QUPC52. The two antibodies also differ in their ability to precipitate a synthetic linear dextran with about 200 glucoses. QUPC52 but not W3129 forms precipitins in saline with the linear sugar chain of α(1,6)dextran. These observations were interpreted to indicate that the combining sites of QUPC52 must be "grooves" where internal chain epitopes of α(1,6)linked glucose could fit. However, the binding sites of W3129 that hold the non-reducing ends of dextran were termed as the "cavity" type. As shown in figure 1, branching present in a dextran molecule can alter the ratio of available terminal epitopes and internal chain epitopes. Computer model building studies of the two cavity-type mAb, W3129 and 16.4.12E, and one groove-type mAb 19.1.2 support the immunochemical distinction of these two basic types of combining-sites (*41, 42*). Recognition of either terminal moieties or internal structures of a carbohydrate molecule may also occur in other protein-carbohydrate interactions.

The solution behavior of carbohydrates, such as the expression of both terminal and internal chain epitopes, reflects a physicochemical property of a carbohydrate molecule. Carbohydrate molecules are rich in hydroxyl groups, and in an aqueous solution, their hydroxyl groups readily interact with water molecules via hydrogen bonding. Since the glycosidic linkages are more flexible than the peptide bonds in proteins,

Terminal and Internal Chain Epitopes of $\alpha(1,6)$dextran

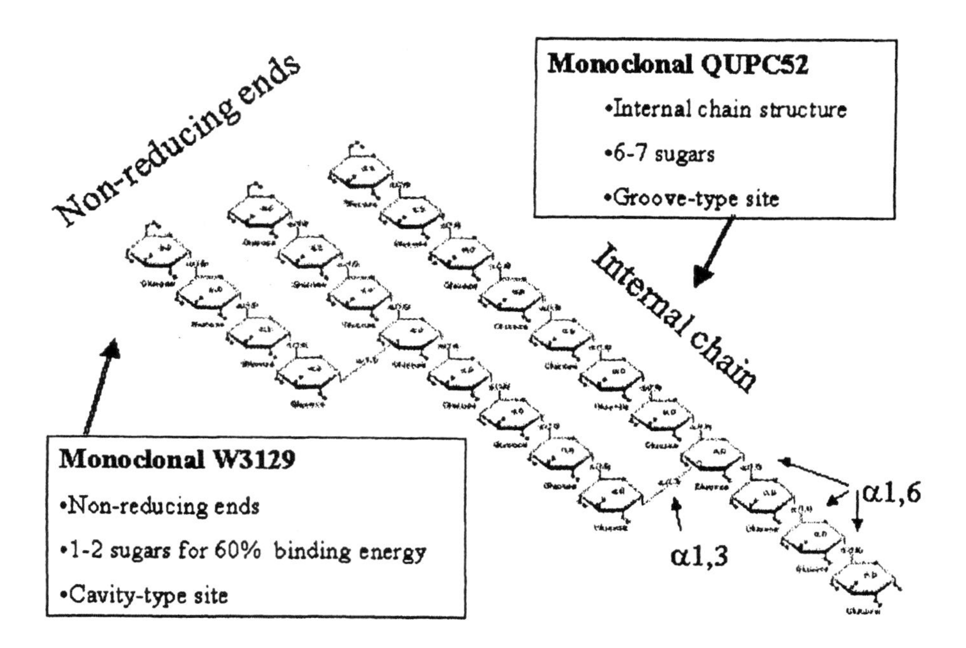

Figure 1: Two types of carbohydrate-based antigenic determinants. The diagram illustrates a schematic view of a portion of the polysaccharide α(1→6) dextran N279 and the two different specificities of anti-α(1→6) dextran antibodies. Antibodies bound to this polysaccharide can be either cavity-type (W3129) for terminal epitopes or groove-type (QUPC52) for internal epitopes. The size of the antibody-binding site can be determined by the competitive inhibition of the dextran-antibody interaction using oliogsaccharides of different numbers of sugar residues.

the protein-like folding patterns, in which the hydrophobic side-chains form an oily core with the polar side-chains exposed, are not observed in polysaccharides. Instead, not only are the terminals of the carbohydrates accessible for molecular recognition, but the residues in the internal chains are also exposed in solvent and are reactive with proteins in solution.

A carbohydrate-based microarray technology

We recently reported the establishment of a high throughput biochip platform for producing a carbohydrate microarray (*2*). This technology takes advantage of state-of-the-art technologies such as the cDNA microarray in terms of both software and hardware. A high-precision robot designed to produce cDNA microarrays was utilized to spot carbohydrate and protein antigens onto glass slides pre-coated with nitrocellulose polymer. The capacity of the slide is approximately 20,000 spots per slide, with 150 picoliters of sample per spot. The antibody-stained slides were then scanned for fluorescent signals with a Biochip Scanner that was developed for cDNA microarrays.

We found that polysaccharides and various glycoconjugates of distinct structural characteristics can be immobilized on a nitrocellulose-coated glass microslide and that their immunological properties can be preserved on chips for long periods of time. In our initial experiments, a well-studied model system, α(1,6)dextran and anti-α(1,6)dextran antibodies, were applied. To investigate whether immobilized carbohydrate macromolecules preserve their antigenic determinants, dextran preparations of different linkage compositions and with different ratios of terminal to internal epitopes were printed on nitrocellulose-coated glass slides. These preparations included N279, displaying both internal linear and terminal nonreducing end epitopes, B1299S, heavily branched and expressing predominantly terminal epitopes, and LD7, a synthetic dextran composed of 100% α(1,6)-linked internal linear chain structure. The dextran microarrays were incubated with monoclonal antibodies of defined specificities, either a groove-type anti- α(1,6)dextran 4.3F1 (IgG3) (*31*) or a cavity-type anti- α(1,6)dextran 16.4.12E (IgA) (*43*). The former recognizes the internal linear chain of α(1,6)dextrans; while the latter is specific for the terminal nonreducing end structure of the polysaccharide.

The groove-type mAb, 4.3F1, bound well to the dextran preparations with predominantly linear chain structures, N279 and LD7, but bound poorly to the heavily branched α(1.6)dextran, B1299S. By contrast, when the cavity-type mAb 16.4.12E was applied, it bound to the immobilized dextran preparations with branches (N279 and B1299S), but not to those with only internal linear chain structure (LD7). These patterns of antigen-antibody reactivities are characteristically identical to those recognized by an ELISA binding assay, and other classical quantitative immunoassays for either the groove-type or cavity type of anti-dextran mAbs (*2*). Therefore, the immunological properties of dextran molecules are well preserved when immobilized on a nitrocellulose-coated glass slide. Their non-reducing end structure, recognized by the cavity-type anti-α(1,6)dextrans, as well as the internal linear chain epitopes, bound by the groove-type anti-α(1,6)dextrans, are displayed on the surface after immobilization and are accessible to antibodies in an aqueous solution.

A high potential biochip platform

We have demonstrated that this microarray platform is also applicable for producing protein microarrays. As with carbohydrate microarrays (*2*), there is no need to chemically conjugate or denature a protein of sufficient molecular weight for its surface immobilization. Protein preparations, including antibodies and protein antigens of various origins can be stably immobilized on the surface. As seen in Figure 2 below, a number of glycoproteins were arrayed on a biochip. The chip locations of these protein preparations are indicated in the figure legend.

In addition, a number of artificial glycoconjugates, i.e., isomaltosyl oligosaccharide-protein conjugates of IM3-BSA, IM6-BSA, IM3-KLH and IM6-KLH were applied onto the chip to display their oligosaccharides. In figure 3, they were printed at chip locations E1, E2, M1, and M2, respectively. Except for IM6-KLH (M2), which was barely detectable by all four-antibody preparations used for screening, other glycoconjugates displayed their oligosaccharide chains on the chip. The preparation of IM6-KLH was confirmed by other immunoassays as a bad stock and was removed from our chips in the later experiments. Thus, to summarize, we also demonstrated in this publication (*2*) a method for the production of oligosaccharide microarrays.

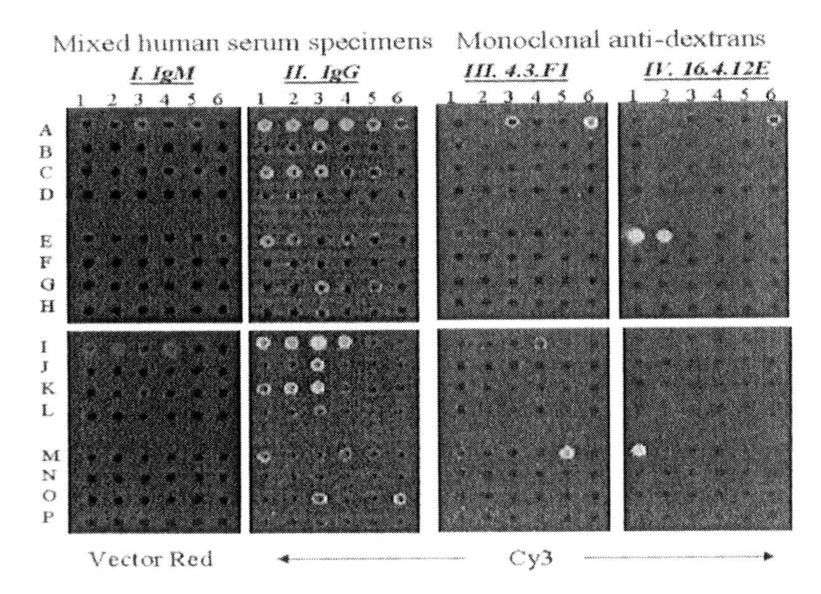

Figure 2. An antigen microarray for detection and characterization of human and murine antibodies with anti-carbohydrate activities. Forty-eight distinct antigen preparations were arrayed on slides at antigen concentrations of 0.5 mg/ml and 0.02 mg/ml. Their antibody reactivities after surface-immobilization were tested with mixed human and murine antibodies. The human IgM captured by microarrays was visualized using an anti-human IgM-AP conjugate and the color developed using Vector Red. The human IgG anti-carbohydrates were detected using a biotinylated anti-human IgG and visualized by a Cy3-Streptoavidin. Preparations of natural and semi-synthetic glycoproteins printed on this chip were C4 (array location)-Cow 21(Blood group B); E1-IM3-BSA; E2- IM3-KLH; E3-Le^{a} (N-1 10% 2X); E4-Beach P1 (Blood group B); E5-Tij II (Blood group B and Le^{a}); E6-OG; G1-ASOR; G2-LNT-BSA; K4-Cow26 (Blood group B); M1-IM6-BSA; M2-IM6-KLH; M3-Le^{a} (N-1 IO-4 NaOH); M4-Cyst 9 (Blood group A); M6-Hog (Blood Group H); and O1-AGOR. (Reprint of Wang et al. Nature Biotechnology, 20:275, 2002. Permission granted by *Nature Biotechnology*.) (*2*)

Lastly, we found that a groove-type anti-dextran mAb 4.3.F1, which is complementary to a sugar chain of six sugar residues, was unable to bind an IM6-BSA conjugate (figure 2). In this assay, two anti-dextran mAbs of well-characterized combining-site, 4.3.F1 (Groove-type) and 16.4.12E (Cavity-type), were applied to interact with a large panel of carbohydrate antigens on chip. 16.4.12E (cavity-type) but not 4.3.F1 (groove-type) bound to isomaltotriose-coupled BSA (IM3-BSA) and IM3-KLH (Fig. 3). Similarly, 4.3.F1 (Groove-type) but not 16.4.12E binds to a linear dextran N-150-N. These results confirm the epitope-specific binding specificities of the two monoclonal antibodies. It was, however, interesting to see that 16.4.12E (Cavity-type) but not 4.3F1 (Groove-type) was able to bind IM6-BSA. Although the groove-type anti-α(1,6)dextran, 4.3F1, has combining sites complementary to the isomaltohexaose (IM6), as shown by oligosaccharide competition assays, it was not able to react with IM6-BSA. This provides evidence that the oligosaccharide chain displayed by this glyocojugate is conformationally different from the internal chain epitopes of α(1,6)dextran N279 and therefore was not recognized by mAb 4.3F1.

Conformational flexibility is an intrinsic property of carbohydrate molecules (*14*). In oligosaccharides and polysaccharides, all component monosaccharides exist in the ring configuration, except for the terminal-reducing residue, which is at equilibrium between its ring and its open-chain form. The geometry of the individual sugar rings in a sugar chain is essentially rigid. However, adjacent monosaccharides can be rotated around their glycosidic bonds. Thus, a sugar chain may exist in multiple conformations. Recent studies on the capsular polysaccharide of *group B meningococcus* found that this polysaccharide can adopt a unique antigenic conformation of helical structures, which are pathogen-specific and are not cross-reactive with the smaller oligomers of sialic acid expressed by human tissue (*44, 45*). This pioneered the design and development of chemically modified sugar chain vaccines, in which the pathogen-specific helical structures of polymers were better preserved.

In summary, understanding the nature and characteristics of the antigenic diversity of carbohydrate molecules is critically important for our consideration in establishing carbohydrate microarray technologies. An efficient and successful biochip must take into account the various types of carbohydrate epitopes, such as terminal and internal chain structures, multiple conformers of a given structure, and dominant and minor antigenic determinants. The platforms of oligosaccharide-based biochips may be sufficient to display the terminal chain structures of carbohydrate antigens. Such a platform can be used to study the relatively simple terminal non-reducing end sugar epitopes, as well as branched structures at the termini of complex carbohydrates. The latter type of structures is relatively stable in conformation. We must, however, keep in mind that an oligosaccharide chain of limited number of sugar residues does not necessarily preserve the configurations of the internal chain structures of a polysaccharide. Given the complexity of carbohydrates, specific experiments must be conducted to determine whether a specific oligosaccharide preserves a required antigenic configuration of a polysaccharide chain. Practically, this type of investigation may not be a straightforward process. In order to ensure the quality and antigenic diversity of a diagnostic sugar chip, considerable efforts must be made to study the nature and characteristics of pathogen-specific sugar epitopes.

References:

1. Feizi, T. Glycoconj. J., 2000. **17**(7-9): p. 553.
2. Wang, D.; Liu, S.; Trummer, B.J.; Deng, C. Wang, A. Nat Biotechnol, 2002. **20**(3): p. 275.
3. Rosati, F.; Capone, A.; Giovampaola, C.D.; Brettoni, C. Focarelli, R. Int. J. Dev. Biol., 2000. **44**(6): p. 609.
4. Focarelli, R.; La Sala, G.B.; Balasini, M. Rosati, F. Cells Tissues Organs, 2001. **168**(1-2): p. 76.
5. Feizi, T. Adv. Exp. Med. Biol., 1982. **152**: p. 167.
6. Crocker, P.R. Feizi, T. Curr. Opin. Struct. Biol., 1996. **6**(5): p. 679.
7. Feizi, T. Trends. Biochem. Sci., 1994. **19**(6): p. 233.
8. Hakomori, S. Cancer. Res., 1985. **45**(6): p. 2405.
9. Sell, S. Hum. Pathol., 1990. **21**(10): p. 1003.
10. Adachi, M.; Hayami, M.; Kashiwagi, N.; Mizuta, T.; Ohta, Y.; Gill, M.J.; Matheson, D.S.; Tamaoki, T.; Shiozawa, C. Hakomori, S. J. Exp. Med., 1988. **167**(2): p. 323.
11. Nakaishi, H.; Sanai, Y.; Shibuya, M.; Iwamori, M. Nagai, Y. Cancer. Res., 1988. **48**(7): p. 1753.
12. Schachter, H. Jaeken, J. Biochim. Biophys. Acta., 1999. **1455**(2-3): p. 179.
13. Mond, J.J.; Lees, A. Snapper, C.M., *T cell-independent antigens type 2*, in *Annual. Review. of. Immunology*. 1995. p. 655.
14. Wang, D. Kabat, E.A., *Carbohydrate Antigens (Polysaccharides)*, in *Structure of Antigens.*, Regenmortal,

M.H.V.V., Editor. 1996, CRC Press: Boca Raton New York London Tokyo. p. 247.

15. Schneerson, R.; Barrera, O.; Sutton, A. Robbins, J.B. J Exp Med, 1980. **152**(2): p. 361.
16. Robbins, J.B. Schneerson, R. J. Infct. Dis., 1990. **161**: p. 821.
17. Ezzell, J.W., Jr.; Abshire, T.G.; Little, S.F.; Lidgerding, B.C. Brown, C. J Clin Microbiol, 1990. **28**(2): p. 223.
18. Cygler, M.; Rose, D.R. Bundle, D.R. Science, 1991. **253**(5018): p. 442.
19. Wuorimaa, T. Kayhty, H. Scand J Immunol, 2002. **56**(2): p. 111.
20. DeRisi, J.; Penland, L.; Brown, P.O.; Bittner, M.L.; Meltzer, P.S.; Ray, M.; Chen, Y.; Su, Y.A. Trent, J.M. Nat Genet, 1996. **14**(4): p. 457.
21. Brown, P.O. Botstein, D. Nature. Genetics., 1999. **21**(1 Suppl): p. 33.
22. Schena, M.; Shalon, D.; Heller, R.; Chai, A.; Brown, P.O. Davis, R.W. Proc Natl Acad Sci U S A, 1996. **93**(20): p. 10614.
23. Ramsay, G. Nature. Biotechnology., 1998. **16**(1): p. 40.
24. Lueking, A.; Horn, M.; Eickhoff, H.; Bussow, K.; Lehrach, H. Walter, G. Analytical. Biochemistry., 1999. **270**(1): p. 103.
25. MacBeath, G. Schreiber, S.L. Science, 2000. **299**(Sept. 8): p. 1760.
26. Stoll, D.; Templin, M.F.; Schrenk, M.; Traub, P.C.; Vohringer, C.F. Joos, T.O. Front Biosci, 2002. **7**: p. C13.
27. Drickamer, K. Taylor, M.E. Genome Biol, 2002. **3**(12).
28. Groenwall, A., *Dextran and its Use in Colloidal Infusion Solutions*. 1957, New York: Academic Press, Inc. 156.
29. Cisar, J.; Kabat, E.A.; Dörner, M.M. Liao, J. J. Exp. Med., 1975. **142**: p. 435.
30. Wood, C. Kabat, E.A. J Exp Med, 1981. **154**(2): p. 432.
31. Wang, D.; Liao, J.; Mitra, D.; Akolkar, P.N.; Gruezo, F. Kabat, E.A. Mol. Immunol., 1991. **28**(12): p. 1387.
32. Kabat, E.A. Berg, D. J.Immunol., 1953. **70**: p. 514.
33. Kabat, E.A. Bezer, A.E. Arch. Biochem. Biophys., 1958. **78**: p. 306.
34. Jeanes, A. Mol. Immunol., 1986. **23**(9): p. 999.
35. Feizi, T. Loveless, R.W. Am. J. Respir. Crit. Care. Med., 1996. **154**(4 Pt 2): p. S133.

36. Karlsson, K.A. Mol. Microbiol., 1998. **29**(1): p. 1.
37. Karlsson, K.A.; Angstrom, J.; Bergstrom, J. Lanne, B. APMIS. Suppl., 1992. **27**: p. 71.
38. Comer, F.I. Hart, G.W. J Biol Chem, 2000. **275**(38): p. 29179.
39. Kamemura, K.; Hayes, B.K.; Comer, F.I. Hart, G.W. J Biol Chem, 2002. **277**(21): p. 19229.
40. Kabat, E.A.; Turino, G.M.; Tarrow, A.B. Maurer, P.H. J. Clin. Invest., 1957. **37**: p. 1160.
41. Padlan, E.A. Kabat, E.A. Proc Natl Acad Sci U S A, 1988. **85**(18): p. 6885.
42. Wang, D.; Hubbard, J.M. Kabat, E.A. J. Biol. Chem., 1993. **268**(27): p. 20584.
43. Matsuda, T. Kabat, E.A. J. Immunol., 1989. **142**: p. 863.
44. Patenaude, S.I.; MacKenzie, C.R.; Bilous, D.; To, R.J.; Ryan, S.E.; Young, N.M. Evans, S.V. Acta Crystallogr D Biol Crystallogr, 1998. **54**(Pt 6 Pt 2): p. 1456.
45. Patenaude, S.I.; Vijay, S.M.; Yang, Q.L.; Jennings, H.J. Evans, S.V. Acta Crystallogr D Biol Crystallogr, 1998. **54**(Pt 5): p. 1005.

Chapter 5

Chemoenzymatic Synthesis of Lactosamine and (α2→3)Sialylated Lactosamine Building Blocks

Fengyang Yan, Seema Mehta, Eva Eichler, Warren Wakarchuk, and Dennis M. Whitfield*

Institute for Biological Sciences, National Research Council of Canada, 100 Sussex Drive, Ottawa, Ontario K1A 0R6, Canada

Recent progress in the development of chemoenzymatic synthesis of oligosaccharide building blocks including the recent success with N-trichloroethoxycarbonyl(Troc) protected glucosamine derivatives to synthesize lactosamine and sialylated lactosamine building blocks is described. The methods outlined here can be scaled to industrial quantities allowing for efficient optimization of glycoconjugate synthesis and activity. Examples include oligosaccharides related to glycolipids, glycoproteins and bacterial capsular polysaccharides.

Introduction

As glycobiology moves from the research laboratory to the industrial world it has become necessary to develop simpler synthetic methods for complex oligosaccharides (*1*). Traditional solution synthesis relies on tedious protection-deprotection strategies coupled with capricious glycosylation reactions all of which almost invariably require time consuming chromatographic separations at every step. Significant amelioration of the separation problem can be achieved by tagging procedures, notably solid phase or more generally polymer-supported methodologies (*2*). These tagging techniques still rely on protection-deprotection chemistry to prepare monosaccharide or in general oligosaccharide building blocks

Published 2004 American Chemical Society

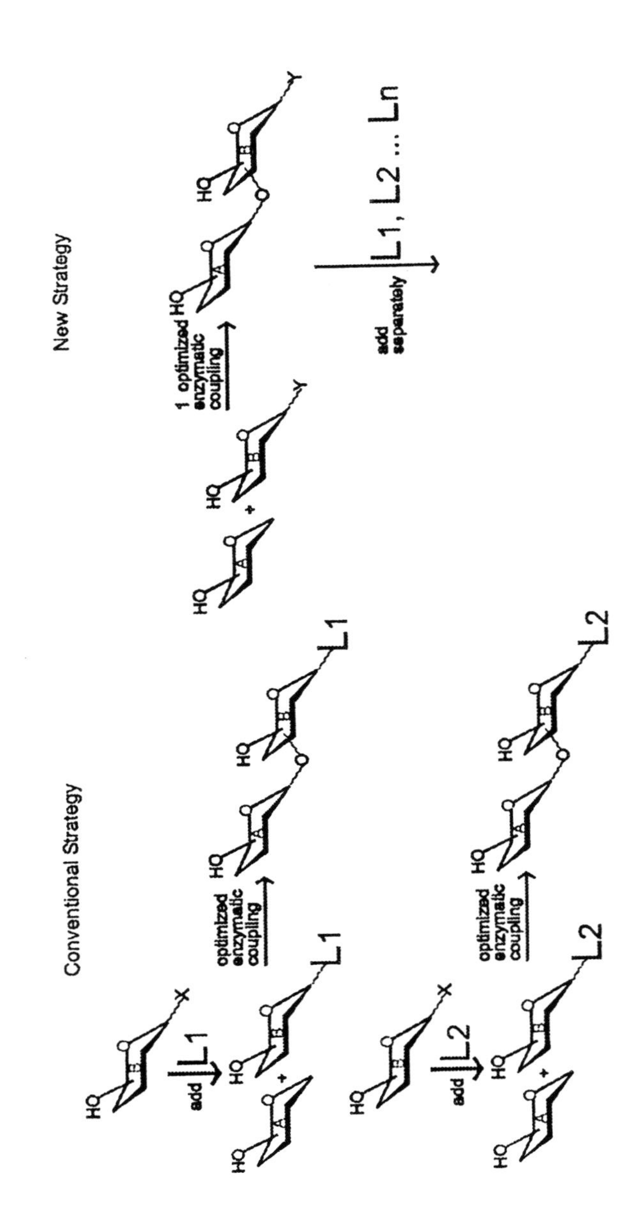
Conventional Strategy
add
L1
optimized enzymatic coupling
L1
add
L2
optimized enzymatic coupling
L2
New Strategy
1 optimized enzymatic coupling
add separately
L1, L2 ... Ln

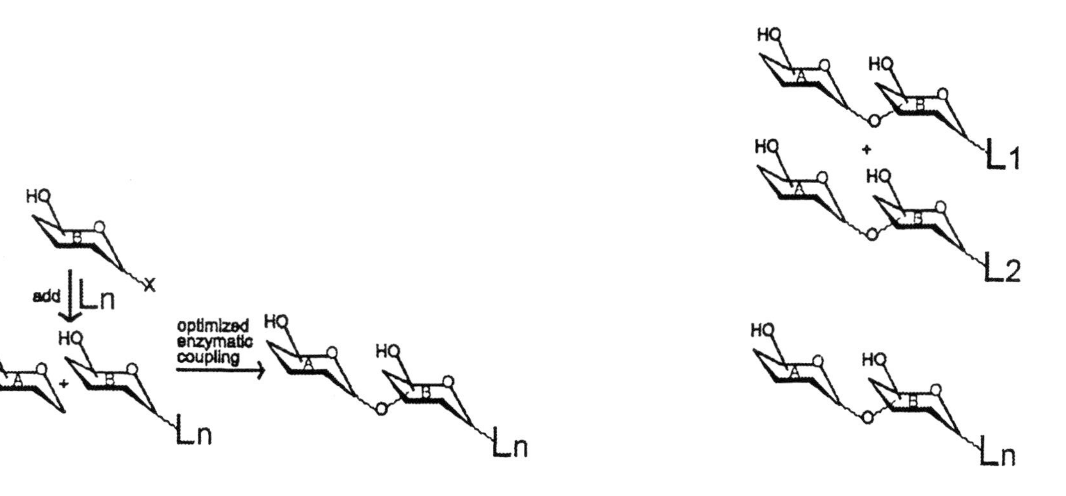

Scheme 1. Optimization of oligosaccharide linking by conventional one at a time linkage plus enzyme optimization or by convergent strategy of optimizing one enzyme reaction and then coupling linkers.

(*3*). For the assembly of larger oligosaccharides most synthetic routes prepare di- tri- and higher oligomeric building block. Often these building blocks contain oligosaccharides with *cis* linkages or other "difficult" linkages such as those with 2-keto-acids like sialic acids that frequently can only be formed in low yield (*4*). Such reactions often lead to complex mixtures that require careful separations. To join these oligosaccharide building blocks high yielding glycosylation reactions that form *trans* linkages via neighboring group participation are typically used (*5*). For industrial applications simple synthetic methods to such oligosaccharide building blocks are highly desirable. This is the driving force behind our research program.

The impending industrialization of glycobiology has also led to another important realization which is that most commercial products will be glycoconjugates and not reducing oligosaccharides. Not only are glycoconjugates more stable but they allow for control of multivalency and the combining of multiple activities. The "classic" example is glycoconjugate vaccines where in almost all cases conjugation of the immunogenic oligosaccharide to a carrier, typically but not necessarily a protein, leads to higher response levels and the desired immunological phenotype (*6*). Especially in the development stage synthetic procedures that produce oligosaccharide building blocks that can be coupled to a variety of linkers and carriers will be more useful than those that require a separate synthesis for each linker-carrier combination. The building block approach allows one route to be optimized instead of having to optimize each route separately. These two strategies are compared in Scheme 1. As outlined by numerous groups in this symposium, glycosyltransferases solve the regiochemical and stereochemical complexities of oligosaccharide synthesis (7). With the recent emergence of a wide variety of cloned and expressed transferases from bacterial sources many enzymes are now available in synthetically useful quantities (*8*). Similarly, following a variety of strategies outlined by other groups in this symposium, the requisite nucleotide donors are also becoming available in bulk at competitive prices (*9*). Our presentation briefly reviews our efforts at using glycosyltransferases to prepare oligosaccharide building blocks including our recent work on suitably N-protected lactosamine, Gal(β1→4)GlcNAc, and (α2→3)sialylatedlactosamine building blocks.

Background

Our group has devoted considerable effort towards developing polymer-supported synthetic methodologies for glycoconjugates of the Group B *streptococcus* type 1A capsular polysaccharide, see Scheme 2. In particular we wish to prepare multiple repeat units of the pentasaccharide repeat unit in a form suitable for conjugation to protein carriers since such constructs are known to elicit protective antibodies (*10*). Our reasons for choosing this target are three fold. First, this is a pathogen causing considerable mortality and morbidity to especially

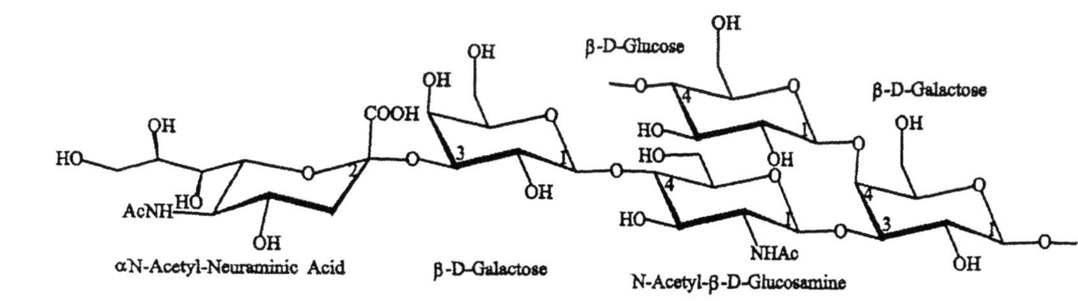

Scheme 2. Chemical structure of a single repeat unit of group B Streptococcus Type 1A capsular polysaccharide.

Protected Trisaccharide

1

UDP-Glc ⇌ UDP-Gal

GalE-LgtB

HEPES, $MnCl_2$, 37 °C

2

CTP + Neu5Ac → CMP-Neu5Ac

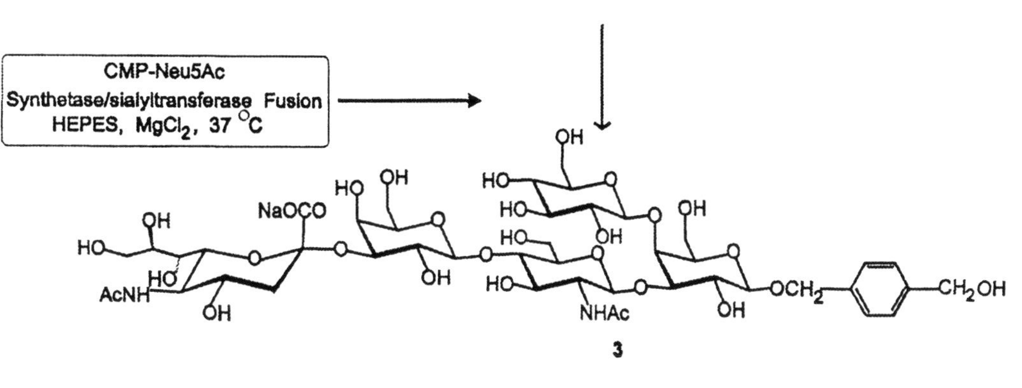

*Scheme 3. Transformation of trisaccharide **1** into pentasaccharide **3** by the use of two fusion enzymes.*

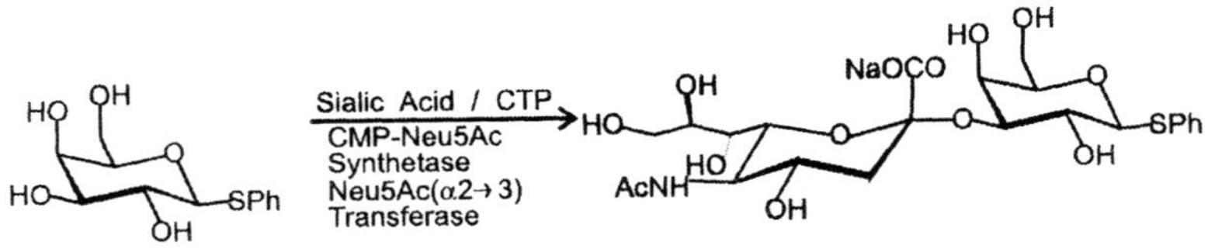

Scheme 4. Chemoenzymatic synthesis of sialylated galactose disaccharide building block.

neonates. Second, a considerable amount of knowledge has been acquired, partly from our Institute, on the immunology of the capsular polysaccharide that can simplify our biological studies. Third, since it is a polymer of β1→4 linked lactose, Gal(β1→4)Glc, with (α2→3)sialylatedlactosamine sidechains and since these oligosaccharides are ubiquitous in biologically relevant targets it is anticipated that the building blocks and synthetic methodologies could have widespread applications (*11*).

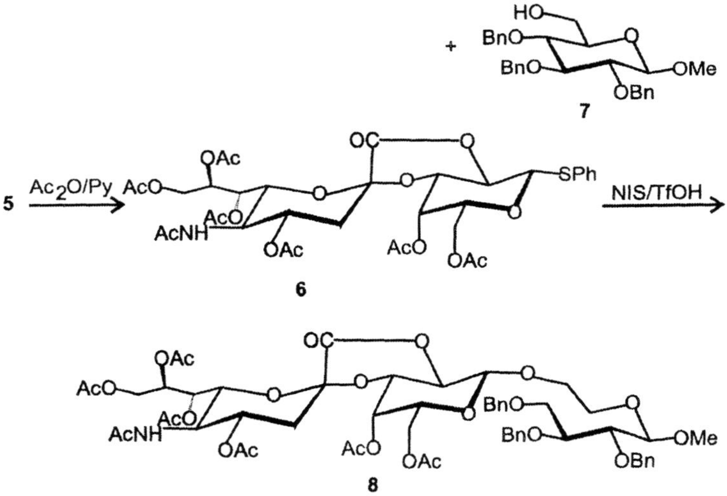

*Scheme 5. Synthesis of β-linked trisaccharide **8** from disaccharide donor **6**.*

To this end we have developed a number of syntheses of protected versions of the branched trisaccharide [GlcNAc(β1→3)]Glc(β1→4)Gal(β1→O)DOXOH, **1**. DOX is an acronym for the dioxyxylene linker (*12*) and its presence in the polymer cleaved product is a result of the Scandium triflate cleavage chemistry developed during this project (*13*). The benzylic position is envisaged to be a reactive site for further linker chemistry. The protecting groups and glycosylation methodologies have been carefully chosen to allow for synthesis of multiple repeat units. A single repeat unit has been synthesized from **1** by using two fusion enzymes (*14*). The first is a GalE/GalT fusion enzyme that allows UDP-Glc to be used as donor as the *in situ* generated UDP-Gal from the GalE activity which is then used to form a Gal(β1→4)GlcNAc linkage due to the GalT(LgtB) activity, cf. **2** in Scheme 3. Subsequently, a CMP-synthetase/ST3 fusion enzyme synthesizes CMP-Neu5Ac from free sialic acid and CTP which is then transferred to the tetrasaccharide to form a Neu5Ac(α2→3)Gal linkage, cf. **3** (*15*). This synthesis is adequate for making a single repeat unit but for multiple repeat units difficulties can be expected in ensuring that each enzyme reaction goes to completion and in separating any incompletely reacted material from the product.

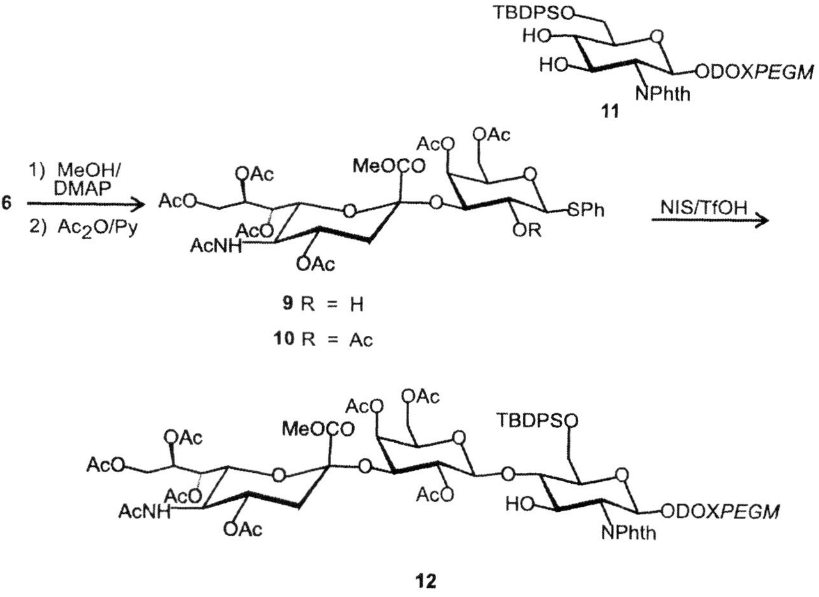

*Scheme 6. Synthesis of sialylated lactosamine derivative **12** from disaccharide donor **10** and polymer-supported acceptor **11**.*

The proposed strategy is to make either lactosamine building blocks or sialylated lactosamine building blocks that can be used instead of the monosaccharide GlcNAc building blocks used above. Simultaneous to this work we have synthesized other building blocks by chemoenzymatic methodologies and the insights from these studies were useful for this project. Our first target was a Neu5Ac(α2→3)Gal building block. There are a large number of solution syntheses of such donors all of which require at least 10 synthetic steps from free sugars and

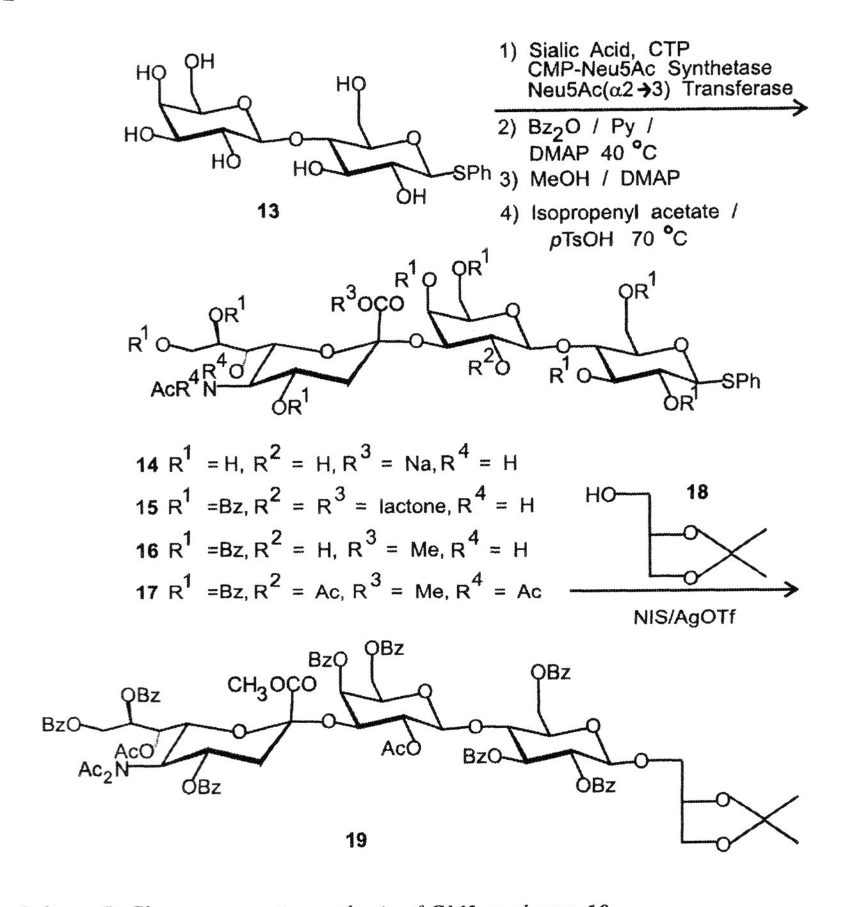

*Scheme 7. Chemoenzymatic synthesis of GM3 analogue **19**.*

multiple column chromatographies (*16*). We found that arylthio glycosides of galactose (for example **4**) are sufficiently good substrates for a variety of bacterial sialyltransferases that essentially quantitative yields of Neu5Ac(α2→3)Gal(β1→SAr) disaccharides such as **5** could be obtained, see Scheme 4 (*17*). Simple acetylation with pyridine/acetic anhydride led to a high yield of lactone **6** that can be purified by conventional silica gel chromatography to provide gram quantities. In principle this five step procedure should be scalable to industrial quantities. Lactone **6** led to high yields of β-linked oligosaccharides such as **8** from alcohol **7**, see Scheme 5. The high stereoselectivity is surprising considering that the lactone should not be a neighboring group. However, the reactivity of this donor is not high. To make a prototype of a sialylatedlactosamine, **6** was opened to the methyl ester **9** which was esterified to give **10**, see Scheme 6. Note that a variety of acyl groups could be introduced at Gal O-2 at this point to control the glycosylation reactivity. Then, **10** was reacted with polymer-supported **11** to yield a protected sialylatedlactosamine derivative **12**. Trisaccharide **12** has an N-phthalimido nitrogen protecting group and the question of the reactivity of these donors will be returned to.

In order to make trisaccharide donors suitable for making ganglioside GM3, cf. Ref. (*18*), mimetics chemoenzymatic sialylation of arylthio glycosides of lactose, for example **14** from **13**, were investigated, see Scheme 7. Since it is well known that 2-O-acetylated lactose derivatives often lead to orthoesters and acyl transfer side reactions (*19*) trisaccharide **14** was benzoylated to give hydroxy-lactone **15**. Subsequent opening to the ester **16** and N– and O-acetylation led to reactive donor **17**. Di-N-acetylation of sialic acid derivatives is known to increase reactivity possibly by minimizing unproductive Lewis acid complexation at the monoamide nitrogen (*20*). Trisaccharide **17** can be used to make GM3 mimetics such as **19** formed by glycosylation of glycerol derivative **18** (*21*).

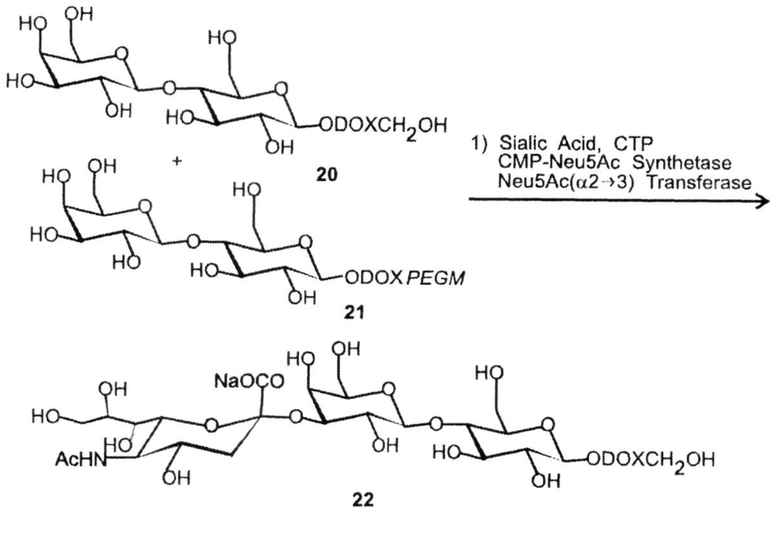

*Scheme 8. Competition experiment between soluble disaccharide **20** and polymer-supported **21** but only **20** reacts.*

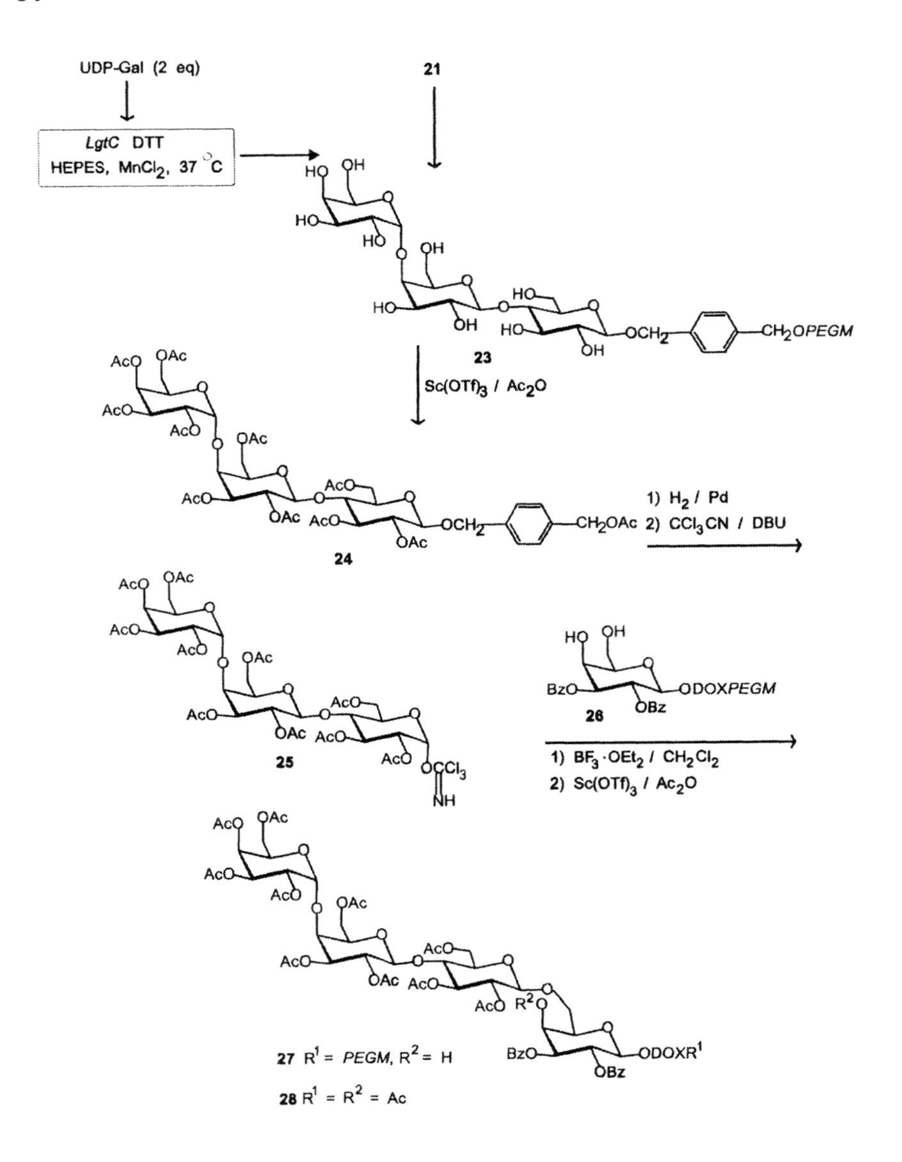

Scheme 9. Chemoenzymatic iterative synthesis of difficult linkages of oligosaccharides on soluble polymeric supports.

One of the problems alluded to above is the purification of products of enzyme reactions. All derivatives discussed above were sufficiently hydrophobic to stick to C-18 reverse phase columns and thus milligram to gram quantities are easily purified. One attractive approach is to attach the acceptor to a polymeric support and then one has to purify the polymer and not the oligosaccharide. The polymeric support we use is the highly water soluble polyethylene glycol, typically we use MPEG-5000, MeO-$(CH_2CH_2O)_{n about 110}$OH. During our studies we have investigated a large number of PEG bound acceptors including lactosamine bound acceptors but with a large number of inverting transferases under a variety of conditions we have found no synthetically useful conditions. The experiment shown in Scheme 8 where lactose attached to the DOX linker **20** was reacted in the presence of a lactose-DOX-PEGM **21** construct to give the GM3 mimetic trisaccharide **22** in high yield and with kinetics comparable to those in the absence of PEG suggests that **21** is neither a substrate or an inhibitor. A reasonable explanation of these results is not apparent to us and requires further investigation. To date two retaining transferases have been reacted with **21** and in both cases a nearly quantitative yield of polymer bound trisaccharides were obtained, cf. **23** in Scheme 9. As shown in Scheme 9 this provides a prototype of a powerful strategy to incorporate *cis* linkages into polymer-supported synthetic schemes (*22*). Thus **23** is readily converted into a trisaccharide

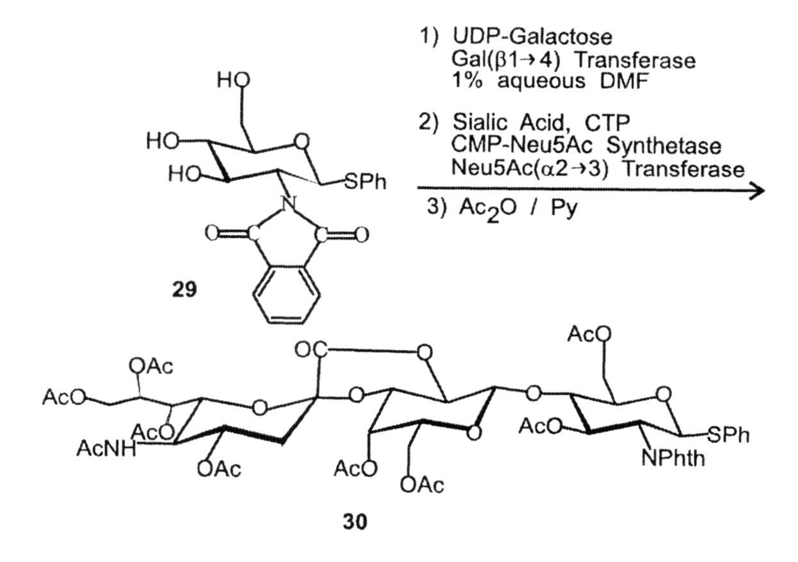

Scheme 10. Chemoenzymatic synthesis of sialylated lactosamine with N-phthalimide as N-protecting group.

donor **25** via the hemi-acetal resulting from hydrogenation of trisaccharide **24** obtained by cleavage from the polymer. Donor **25** can be reacted with polymer bound acceptor **26** to yield tetrasaccharide **27** and after cleavage **28**. Tetrasaccharide **28** contains the *cis* linkage made stereospecifically and regiospecifically by the transferase.

Based on our successful use of monosaccharide donors with N-phthaloyl protection to make trisaccharide **1** and our anticipation that this protecting group should be stable to both the enzymatic and chemical reaction conditions (*23*) we started a program to prepare N-phthaloyl protected lactosamine and (α2→3)sialylatedlactosamine building blocks. All attempts to use PEG bound substrates failed as discussed above. Eventually it was found that the thiophenyl glycoside of N-phthaloyl glucosamine **29** (*24*) solubilized with DMF was a substrate for the GalT and ST enzymes and after acetylation was turned into potential donor **30**, see Scheme 10 (*25*). But, the overall yields were always low even though TLC monitoring of the enzyme reactions suggested that all the substrates were consumed. The problem was traced via NMR and MS analysis of the product mixtures to ring opening of the N-phthalimide in water, see Scheme 11. The use of *p*-methoxybenzyl or even better *p*-nitrobenzyl glycosides of N-phthaloyl glucosamine minimized but did not eliminate this side reaction. With the small quantities of **30** available, several glycosylations with polymer bound acceptors, that worked well with phenyl 3,4,6-tri-*O*-acetyl-2-deoxy-2-phthalimido-β-D-thioglucopyranoside as donor, failed or gave only low yields. Therefore, we decided to investigate other nitrogen protecting groups.

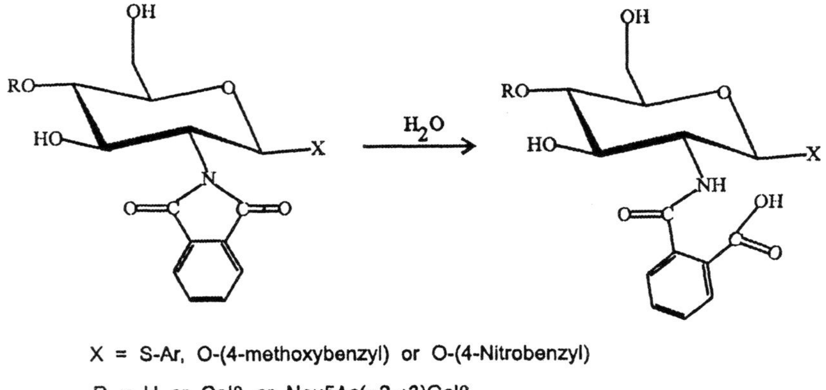

X = S-Ar, O-(4-methoxybenzyl) or O-(4-Nitrobenzyl)

R = H or Galβ or Neu5Ac(α2→3)Galβ

Scheme 11. Ubiquitous ring opening in water of N-phthalimide protecting groups.

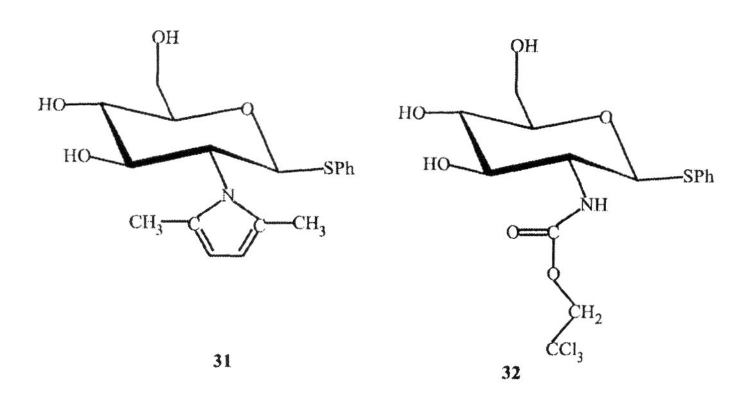

Scheme 12. 2,5-Dimethylpyrrole and 2,2',2''-trichloroethoxycarbonyl(Troc) protecting groups for enzymatic synthesis of lactosamine.

Among the groups studied, two led to good yields of sialylated lactosamine derivatives namely 2,5-dimethylpyrrole (*26*) and trichloroethoxycarbonyl, Troc (*27*), cf. **31** and **32** Scheme 12. In our hands the donors prepared from **31** were not reactive with low nucleophilicity acceptors whereas **33** and **34** derived from **32** (*28*)

32
1) *GalE-LgtB*
HEPES, $MnCl_2$, 37 °C
2) $Sc(OTf)_3$ / Ac_2O
AcO AcO AcO AcO AcO AcO SPh NHTroc
33

1) Sialic Acid, CTP
CMP-Neu5Ac Synthetase
Neu5Ac(α2→3) Transferase
2) CH_3OH / $ResinH^+$
3) $Sc(OTf)_3$ / Ac_2O
OAc AcO AcNH AcO OAc $OCOCH_3$ AcO OAc OAc AcO AcO SPh NHTroc
34

*Scheme 13. Chemoenzymatic synthesis of N-Troc protected sialylated lactosamine donor **34**. CMP-Neu5Ac synthetase (29) and Neu5Ac(α2→3)transferase (30) are known enzymes.*

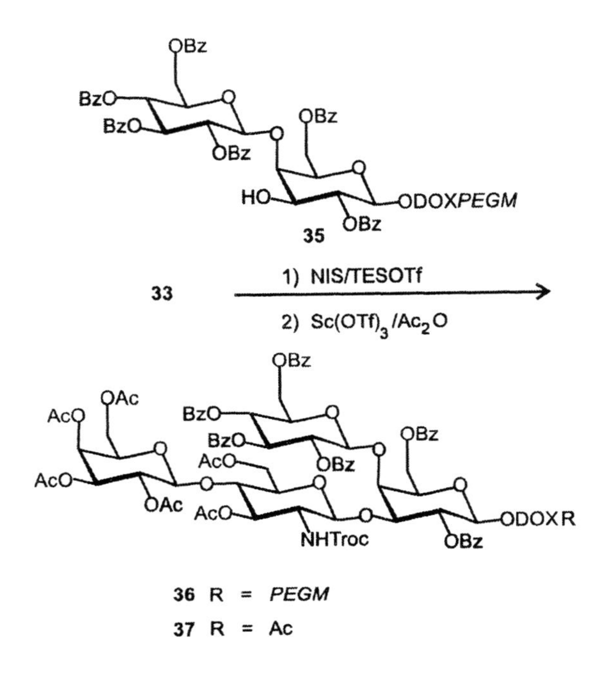

Scheme 14. Synthesis of trisaccharide 37 from unreactive polymer-supported acceptor 35.

were good donors, see Scheme 13. It should be noted that the processing of enzyme product **32** was greatly simplified by using acetic anhydride as the solvent and $Sc(OTf)_3$ as the acetylation catalyst. This process is very simple as the product is isolated by simple aqueous organic solvent extractions followed by standard flash silica gel chromatography. Also, only traces of lactone are formed this way cutting down on the number of synthetic steps from enzyme reaction to useful donor. As an example lactosamine donor **33** was successfully reacted with PEG bound acceptor **35** to yield tetrasaccharide **36** and after $Sc(OTf)_3$ cleavage to yield **37** without affecting the Troc protecting group, see Scheme 14. Similarly the (α2→3)sialylatedlactosamine derived donor **34** could be reacted with mannose OH-2 acceptor **38** to yield tetrasaccharide **39**, see Scheme 15. Standard deprotection of the N-Troc with Zn in acetic acid followed by N-acetylation gave **40**. Subsequently the O-benzyls were removed by hydrogenation and the acyl groups by transesterification followed by hydrolysis to yield **41**. Tetrasaccharide **41** is a model of a typical arm of a N-linked glycopeptide (*31*). We have made donors **33** and **34** on the 100's of milligram scale but there is no reason to expect that multigram quantities are not easily accessible.

Future Prospects

Our goal is to prepare by chemoenzymatic methods a large selection of protected oligosaccharide building blocks that can be used to prepare target glycoconjugates. The building blocks will contain the linkages that are "difficult" to prepare

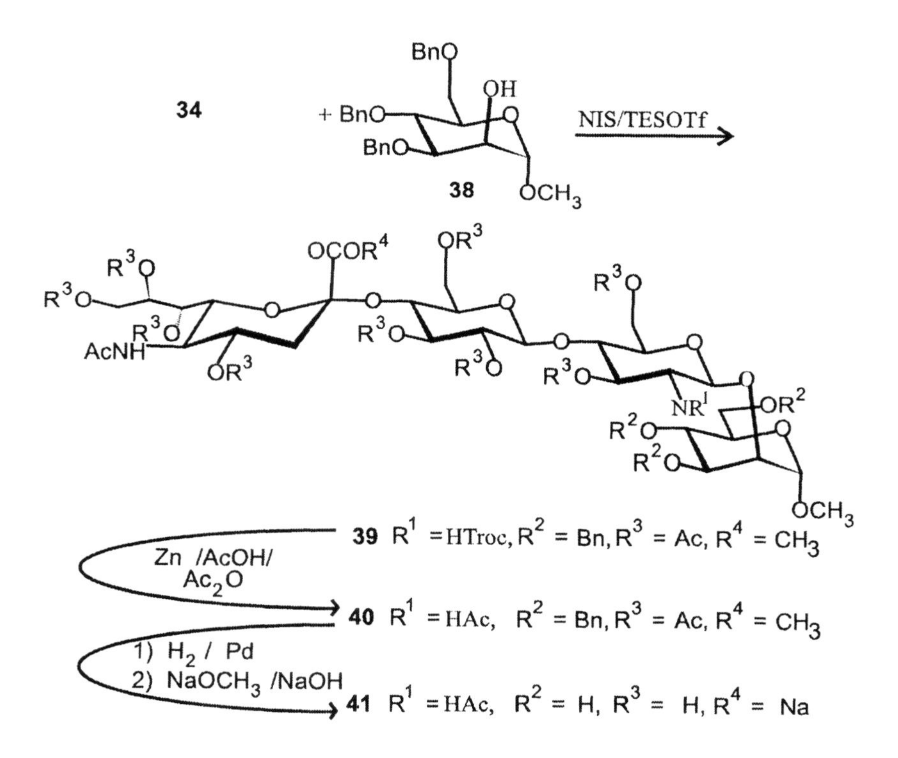

*Scheme 15. Synthesis of tetrasaccharide **41** representative of a N-linked glycopeptide arm using trisaccharide donor **34**.*

chemically with high stereospecificity and in high yield. Instead the required stereoselectivity and high yields will come from optimizing the glycosyltransferase catalyzed reaction. The joining points of the building blocks will be *trans* linkages that can usually be formed chemically in high yield and high stereospecificity via neighboring group participation. Such strategies can be used in the development stage of glycoconjugate therapeutics particularly for optimizing questions of linkers, multivalency and other medicinal chemical parameters. Such a process is shown schematically in Scheme 16. Finally, the block chemical approach described here is highly amenable to process optimization if a particular glycoconjugate is chosen for large scale production. For example, it is likely that careful optimization of the preparation of **36** and its transformation into **39** and **40** could eliminate even more chromatographic separations leading to even higher efficiencies.

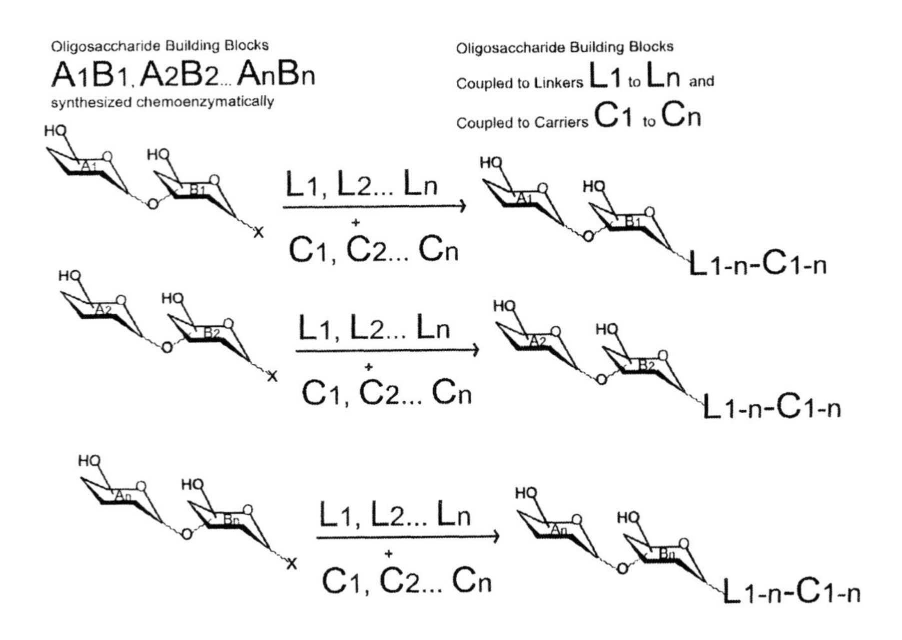

Scheme 16. Directed combinatorial chemistry using chemoenzymatically synthesized oligosaccharide building blocks.

Acknowledgements

This is NRC paper #42468.

References

Abbreviations

GalE, UDP-Glc to UDP-Gal epimerase; GalT, D-galactopyranosyl(β1→4)transferase; ST3, sialyl(α2→3)transferase; CMP, cytidine monophosphate; LgtC, D-galactopyranosyl(α1→4)transferase; DTT, dithiothreitol; DBU, 1,8-diazabicyclo[5.4.0]undec-7-ene; UDP, uridine diphosphate, DMAP, 4-dimethylaminopyridine; pTsOH, *p*-toluenesulfonic acid; HEPES, hydroxyethylpiperazineethylsulfonic acid; TBDPS *t*-butyldiphenylsilyl; py, pyridine; OTf, trifluoromethanesulfonate(triflate); NIS, N-iodosuccinimide; TESOTf, triethylsilyltrifluoromethanesulfonate

1. a) Roth, J. *Chem. Rev.* **2002**, *102*, 285. b) Ritchie, G. E.; Moffatt, B. E.; Sim, R. B.; Morgan, B. P.; Dwek, R. A.; Rudd, P. M. *Chem. Rev.* **2002**, *102*, 305.
2. a) Davis, B. G. *J. Chem. Soc., Perkin Trans. 1*, **2000**, 2137. b) Plante, O. J.; Palmacci, E. R.; Seeberger, P. H. Science, **2001**, *291*, 1523. c) Ye, X.-S.; Wong, C.-H. *J. Org. Chem.* **2000**, *65*, 2410.
3. Boons, G.-J. *Tetrahedron* **1996**, *52*, 1095.
4. Gege, C.; Geyer, A.; Schmidt, R. R. *Chem. Eur. J.* **2002**, *8*, 2454. b) Lergenmiiller, M.; Ito, Y.; Ogawa, T. *Tetrahedron* **1998**, *54*, 1381. c) Turnbull, W. B.; Harrison, J. A.; Kartha, K. P. R.; Schenkman, S.; Field, R. T. Tetrahedron, **2002**, *58*, 3207.
5. a) Hindsgaul, O.; Qian, X. *Chem. Commun.* **1997**, 1059. b) Benakli, K.; Zha, C.; Kerns, R.J. *J. Am. Chem. Soc.* **2001**, *123*, 9461. c) Nukada, T.; Bérces, A.; Whitfield, D.M.; *Carbohydr. Res.* **2002**, *337*, 765.
6. Jennings, H.J.; *Curr. Top. Microbiol. Immunol.* **1990**, *150*, 97.
7. a) Endo, T.; Koizumi, S. *Curr. Opin. Struct. Biol.* **2000**, *10*, 536. b) Wymer, N.; Toone, E. *Curr. Opin. Chem. Biol.* **2000**, *4*, 110. c) Koeller, K. M.; Wong, C.-H. *Glycobiology* **2000**, *10*, 11. d) Palcic, M. M. *Curr. Opin. Biotechnol.* **1999**, *10*, 616.
8. Masayuki, I.; Wong, C.-H. *Trends Glycosci. Glycotechnol.* **2001**, *13*, 345.
9. Bulter, T.; Elling, L. *Glycoconj. J.* **1999**, *16*, 147.
10. Zou, W.; Mackenzie, R.; Thérien, L.; Hirama, T.; Yang, Q.; Gidney, M.A.; Jennings, H.J. *J. Immunol.* **1999**, *163*, 820.
11. a) Dwek, R. A. *Chem. Rev.* **1996**, *96*, 683. b) DeAngelis, P. L. *Glycobiology* **2002**, *12*, 9R.
12. Douglas, S. P.; Whitfield, D. M.; Krepinsky, J. J. *J. Am. Chem. Soc.* **1995**, *117*, 2116.
13. Mehta, S.; Whitfield, D. M. *Tetrahedron* **2000**, *56*, 6415.
14. Eichler, E.; Yan, F.; Sealy, J.; Whitfield, D. M. *Tetrahdron* **2001**, *57*, 6679.
15. Yan, F.; Wakarchuk, W. W.; Gilbert, M.; Richards, J. C.; Whitfield, D. M. *Carbohydr. Res.* **2000**, *328*, 3.

16. a) Ogawa, T.; Nakabayashi, S. *Carbohydr. Res.* **1981**, *97*, 81. b) Kaji, E.; Lichtenthaler, F. W.; Osa, Y.; Zen, S. *Bull. Chem. Soc. Jpn.* **1995**, *68*, 1172.
17. Mehta, S.; Gilbert, M.; Wakarchuk, W. W.; Whitfield, D. M. *Org, Lett.* **2000**, *2*, 751.
18. a) Blixt, O.; Allin, K.; Pereira, L.; Datta, A.; Paulson, J.C. *J. Am. Chem. Soc.* **2002**, *124*, 5739. b) Ito, Y.; Paulson, J.C. *J. Am. Chem. Soc.* **1993**, *115*, 1603.
19. a) Kunz, H.; Harreus, A. *Liebigs Ann. Chem.* **1982**, 41. b) Sato, S.; Nunomura, T.; Ito, Y.; Ogawa, T. *Tetrahedron Lett.* **1988**, *29*, 4097. c) Sato, S.; Ito, Y.; Ogawa, T. *Tetrahedron Lett.* **1988**, *29*, 5267. d) Nunomura, S.; Ogawa, T. *Tetrahedron Lett.* **1988**, *29*, 5681.
20. Demchenko, A.V.; Boons, G.-J. *Chem. Eur. J.* **1999**, *5*, 1278.
21. Acceptor **18** is available from Toronto Research Chemicals.
22. Yan, F.; Gilbert, M.; Wakarchuk, W.W.; Brisson, J.-R.; Whitfield, D.M. *Org. Lett.* **2001**, *3*, 3265.
23. Yan, F.; Mehta, S.; Wakarchuk, W.W.; Gilbert, M.; Schur, M.J.; Whitfield, D.M. *J. Org. Chem.* **2003**, in press.
24. Debenham, J.; Rodebaugh, R.; Fraser-Reid, B. *Liebigs Ann./Recueil*, **1997**, 791.
25. Robina, I.; Gomez-Bujedo, S.; Fernandez-Bolanos, J. G.; Pozp, L. D.; Demange, R.; Picasso, S.; Vogel, P. *Carbohydr. Lett.* **2000**, *3*, 389.
26. Bowers, S. G.; Coe, D. M.; Boons, G.-J. *J. Org. Chem.* **1998**, *63*, 4570.
27. a) Dullenkopf, W.; Castro-Palomino, J.C.; Manzoni, L.; Schmidt, R.R. *Carbohydr. Res.* **1996**, *296*, 135. b) Ellervik, U.; Magnusson, G. *Carbohydr. Res.* **1996**, *280*, 251.
28. Nilsson, K. G. I. *Tetrahedron Lett.* **1997**, *38*, 133.
29. Karwaski, M.-F.; Wakarchuk, W.W.; Gilbert, M. *Protein Expression and Purification*, **2002**, *25*, 237.
30. Gilbert, M.; Bayer, B.; Cunningham, A.-M.; DeFrees, S.; Gao, Y.; Watsor, D. C.; Young, N. M.; Wakarchuk, W. W. *Nature Biotechnol.* **1998**, *16*, 769.
31. a) Sherman, A. A.; Yudina, O. N.; Shashkov, A. S.; Menshov, V. M.; Nifantev, N. E. *Carbohydr. Res.* **2002**, *337*, 451. b) Seifert, J.; Ogawa, T.; Kurono, S.; Ito, Y. *Glycoconjug. J.* **2000**, *17*, 407.

Chapter 6

Synthesis of Bioactive Glycopeptides through Endoglycosidase-Catalyzed Transglycosylation

Lai-Xi Wang, Suddham Singh, and Jiahong Ni

Institute of Human Virology, University of Maryland Biotechnology Institute, University of Maryland, 725 West Lombard Street, Baltimore, MD 21201

The carbohydrate portions of glycoproteins are implicated to play a major role in modulating the structure and biological functions of glycoproteins. However, deciphering the structure and functions of glycoproteins is often hampered by the difficulties in obtaining homogeneous glycoproteins and glycopeptides. This article reviews the transglycosylation with endo-β-*N*-acetylglucosamindases and their application for glycopeptide synthesis. In contrast to glycosyltransferase that adds monosaccharide units one at a time, the endoglycosidase transfers an oligosaccharide moiety to a suitable acceptor in a single step. Incorporation of endoglycosidases into synthetic strategy has made available an array of homogeneous, complex glycopeptides for structural and biological studies.

Glycosylation is a major posttranslational modification of proteins. The oligosaccharide component of glycoproteins plays an important role in many biological processes such as cell adhesion, cell differentiation, host-pathogen interactions, and immune responses (*1-3*). Glycosylation of proteins also affects protein's folding, stability, antigenicity and immunogenicity. However, a clear understanding of the biological functions of glycoproteins is often hampered by the difficulties in obtaining homogeneous glycoproteins or glycopeptides. In fact, most natural or recombinant glycoproteins are heterogeneous and appear as glycoforms, in which a common polypeptide backbone bears varied oligosaccharide chains. Despite tremendous progress in recent years in the art of oligosaccharide and glycopeptide synthesis, the making of homogeneous glycopeptides containing large native oligosaccharide chains is still challenging (*4-8*). It is well accepted that the combination of chemical and enzymatic methods offers the greatest potential in the endeavors (*9-12*). In addition to glycosyltransferases and common glycosidases that have been widely used for oligosaccharide and glycoconjugate synthesis, a special class of hydrolytic enzymes, the endoglycosidases, are emerging as a powerful tool for oligosaccharide and glycoconjugate synthesis. Endoglycosidases are hydrolytic enzymes that cleave the middle glycosidic linkage to release an oligosaccharide moiety from glycoconjugates or polysaccharides. But some of endoglycosidases have demonstrated high transglycosylation activity, the ability to transfer the releasing oligosaccharide moiety to an acceptor other than water to form a new glycosidic bond. These include ceramide glycanase (*13,14*), chitinase (*15*), cellulase (*16*), hyaluronidase (*17*), endo-α-*N*-acetylgalactosaminidase (*18*), and endo-β-*N*-acetylglucosaminidase (*19*). These endoglycosidases have been exploited for chemoenzymatic synthesis of novel polysaccharides and glycoconjugates. This article provides an overview on the transglycosylation of endo-β-*N*-acetylglucosaminidases, with a focus on their application for the synthesis of structurally and biologically interesting glycopeptides.

Transglycosylation Activity of Endo-β-*N*-Acetylglucosaminidases

The endo-β-*N*-acetylglucosamindases (EC 3.2.1.96) are a class of enzymes that hydrolyze the β-1,4-linked glycosidic bond in the *N,N'*-diacetylchitobiose core of the sugar chains in N-glycoproteins to release the oligosaccharide and leave an intact GlcNAc moiety on the asparagine residue in the protein. The enzymes are widely distributed in microorganisms, plants, animals, and humans. Endo-β-*N*-acetylglucosamindases are particularly useful for structural analysis of glycoproteins, facilitating the isolation of intact N-oligosaccharides and partially deglycosylated proteins without degradation. Besides hydrolyzing activity, some endo-β-N-acetylglucosamindases were found to have transglycosylation acitivity, i.e., the ability to transfer the releasing oligosaccharide moiety to a suitable acceptor other than water to form a new glycosidic bond. Among others, the endo-β-*N*-acetylglucosamindase isolated from *Flavobacterium meningosepticum* (Endo-F) was first reported to have a transglycosylation activity capable of transferring the releasing oligosaccharide from a glycopeptide to glycerol (*20*). Next, Takegawa et al (*21-23*) reported that the endo-β-*N*-acetylglucosamindase from *Arthrobacter protophormiae* (Endo-A) could transfer a high-mannose type oligosaccharide to monosaccharides such as N-acetylglucosamine (GlcNAc) and glucose (Glc) to form a new oligosaccharide. Endo-A was found to be specific for high-mannose type oligosaccharides and does not hydrolyze or transfer complex type oligosaccharide chains. On the other hand, Kadowaki, Yamamoto, and co-workers discovered that Endo-M, another endo-β-*N*-acetylglucosamindase isolated from *Mucor hiemalis* could hydrolyze all three types of N-glycans and exhibited significant transglycosylation activity, allowing the transfer of both complex and high-mannose type sugar chains to a suitable acceptor (*24-28*).

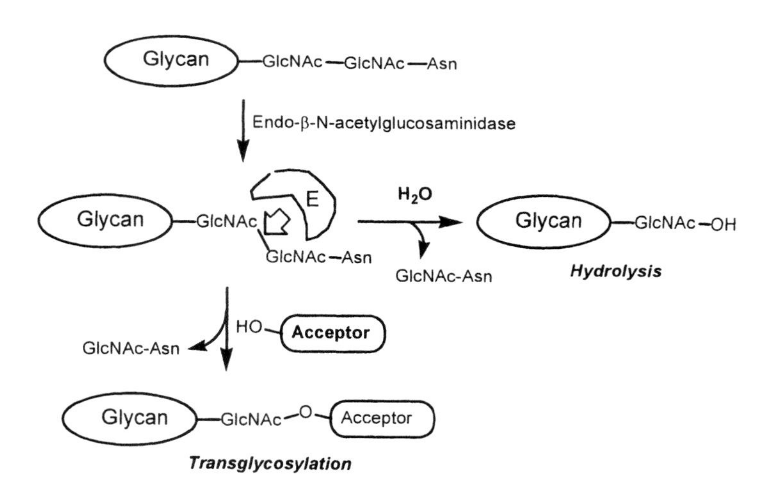

Figure 1. Transglycosylation and hydrolysis with endo-β-N-acetylglucosaminidases

As shown in Figure 1, the enzymatic transglycosylation and hydrolysis are a competing process. The efficiency to form a new oligosaccharide or glycoconjugate by the enzymatic reaction depends on the intrinsic transglycosylation activity of the endoglycosidase and on the control of the reaction conditions (e.g., the concentrations of glycosyl donor and acceptor, and reaction time). Fan et al (*29,30*) found that enhanced transglycosylation could be achieved by adding organic solvents to the reaction media. When the acceptor is a GlcNAc-peptide, a suitable glycopeptide with a native oligosaccharide moiety will be obtained. Thus, a complex type glycopeptide can be synthesized through Endo-M catalyzed transglycosylation, while a high-mannose type glycopeptide can be obtained through either Endo-A or Endo-M catalyzed reaction, when a suitable oligosaccharide donor is used. Recent progress on the endoglycosidase-catalyzed transglycosylation for the synthesis of biologically interesting glycopeptides is described below.

Endoglycosidase-Catalyzed Synthesisof Glycosylated Bioactive Peptides

Glycosylated Calcitonin

Calcitonin is a calcium-regulating peptide hormone that inhibits osteoclastic bone resorption by disrupting actin ring structure of osteoclast cells in a dose-dependent manner (*31*). Eel calcitonin consists of 32 amino acids and exists in a α-helical conformation (*32*). Th peptide has a consensus sequence for N-glycosylation but does not appear glycosylated in its native form. Haneda, Inazu and co-workers developed a method to add a complex or high-mannose type native oligosaccharide to calcitonin through endoglycosidase-catalyzed transglycosylation (*33,34*). First, a calcitonin derivative **1** containing a GlcNAc moiety at the consensus glycosylation site, CT-GlcNAc, was prepared by solid phase peptide fragment synthesis, thioester fragment ligation, and final oxidative cyclization. Transglycosylation of a disialo complex type oligosaccharide to CT-GlcNAc was achieved using the complex type disialo oligosaccharide-asaparagine as the glycosyl donor under the catalysis of Endo-M (Scheme 1). The resulting calcitonin glycopeptide **2** was isolated by HPLC in 8.5% (based on the glycosyl donor used). Endo-M can also transfer a high-mannose type oligosaccharide to CT-GlcNAc to give the high-mannose type calcitonin glycopeptide **3** (CT-M6), but in a low yield (3.5%). However, when Endo-A, which is specific for high-mannose type oligosaccharide, was used for the transglycosylation, CT-M6 was obtained in 32.7% yield under the optimal conditions (*35*) (Scheme 1).

Hashimoto et al (*36*) studied the effects of glycosylation on the structure and dynamics of eel calcitonin in micelles and lipid bilayers by NMR. It was found that the glycosylated calcitonin derivatives, CT-GlcNAc and CT-M6, took the identical comformations in micelles as calcitonin itself. Therefore, the overall conformation of the peptide was not affected by the glycosylation. Similarly,

Tagashira et al (*37*) investigated the conformations of additional calcitonin glycopeptides by CD and NMR and reached the same conclusion. The synthetic calcitonin glycopeptides, CT-GlcNAc and CT-M6, were sufficiently active for disrupting actin ring structure of osteoclast cells, although their activities are lower than that of calcitonin. But when tested in vivo, the CT-GlcNAc that contains a single monosaccharide showed enhanced hypocalcemic activity, while the CT-M6 that bears a high-mannose oligosaccharide moiety exhibited decreased activity in comparison with the elcatonin (*33,34,37*). These studies indicate that glycosylation can modulate the biological activity of calcitonin by attaching distinct carbohydrate structures without altering the peptide backbone conformation.

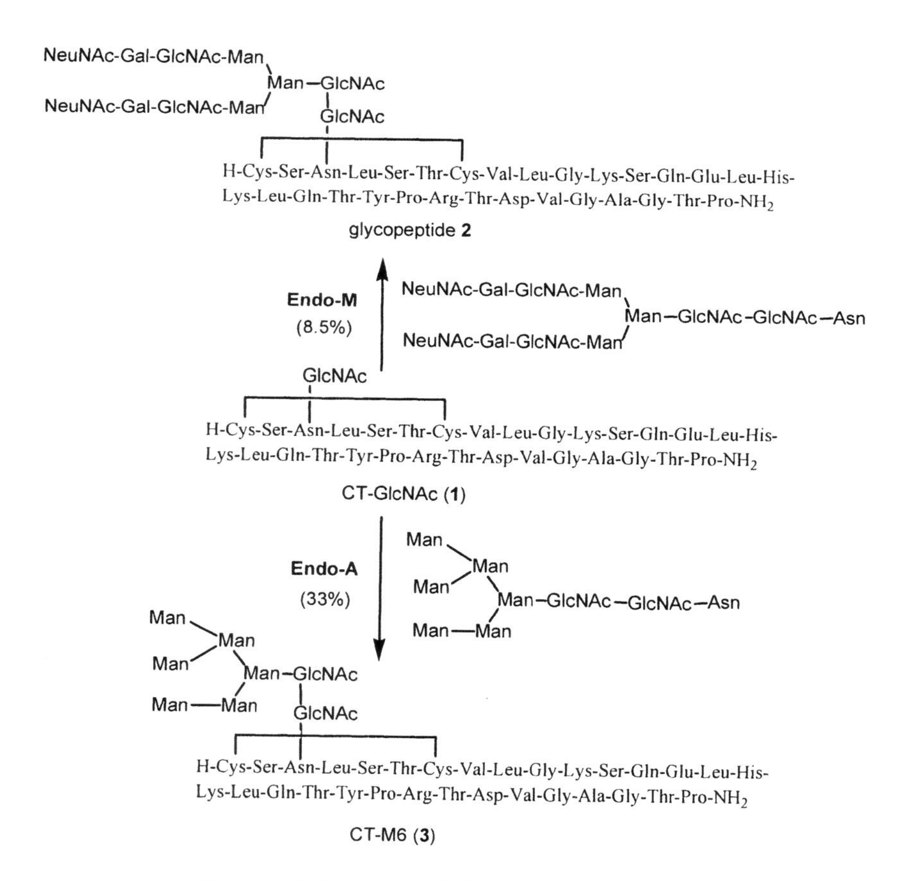

Scheme 1. Synthesis of glycosylated calcitonin

Glycosylated Substance P

Substance P is a neuropeptide that acts on the central and peripheral nervous systems (*38*). The decapeptide hormone contains two L-glutamine residues at the fifth and sixth positions, respectively, but does not have consensus N-glycosylation sequence. Taking the transglycosylation activity of Endo-M, Haneda et al (*39*) artificially added a complex type oligosaccharide chain at the fifth or sixth Gln positions of substance P, giving the SP-5-glycopeptide and SP-6-glycopeptide in 21 and 19.5% yields, respectively (Scheme 2). This study revealed that the Endo-M enzyme could take the non-natural Gln-linked GlcNAc as the glycosyl acceptor equally efficiently, thus allowing the synthesis of non-natural glycopeptides for probing biological problems.

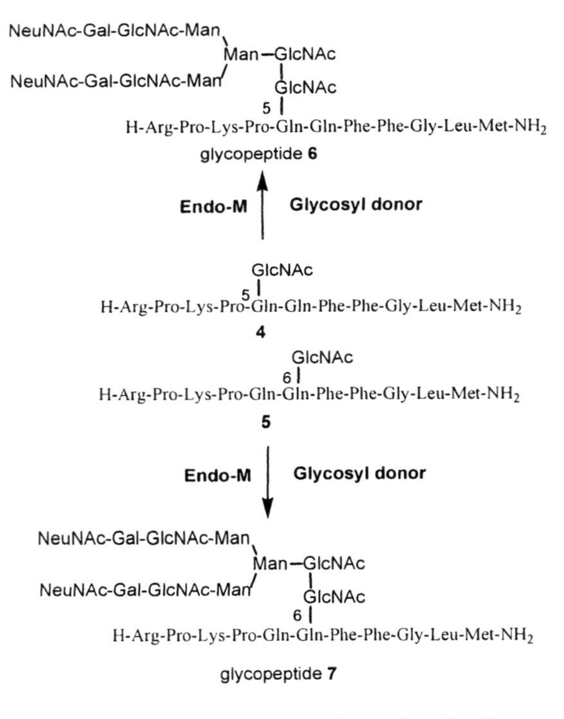

Scheme 2. Synthesis of glycosylated substance P

The biological activity of the synthetic glycosylated substance P derivatives was studied by measuring their effects on muscle contraction of longitudinal strips of guinea pig ileum. The addition of a single GlcNAc at the fifth or the sixth Gln residue did not change the activity of the peptide. However, the attachment of a large complex type oligosaccharide at the sixth Gln residue (compound 7) resulted in a significant loss of its biological activity (*39*). The stability of the glycosylated substance P against proteases such as angiotensin-converting enzyme (ACE) and chymotrypsin was studied. While the native substance P was degraded completely within 5 hours by ACE, the substance P glycopeptides **6** and 7, in which an oligosaccharide moiety was attached to the Gln residue, were totally resistant to ACE and chymotrypsin. The Gln-linked glycopeptides **6** and 7 are also resistant to glycoamidases that otherwise hydrolyze the natural Asn-linked (i. e., N-linked) glycopeptides.

Glycosylated α-Mating Factor and Peptide T

Additional examples were reported for the Endo-M catalyzed glycosylation of bioactive peptides. α-Mating factor is a 13-mer peptide hormone secreted by haloid α cells of yeast *Saccharomyces cerevisiae*. Saskiawan et al (*40*) reported the synthesis of glycosylated α-mating factors **10** and **11**, in which a complex type or a high-mannose type oligosaccharide was attached to a Gln residue in the peptide, respectively (Figure 2). The synthesis also consisted of two key steps: the preparation of the GlcNAc-containing α-mating factor **9** by solid phase peptide synthesis and the transfer of an oligosaccharide moiety to the Gln-linked GlcNAc in the peptide under the catalysis of Endo-M. The biological activity of the synthetic α-mating factor derivatives was analyzed by means of a growth arrest assay using a protease-deficient mutant of α-cells of *S. cerevisiae* (*40*). The large glycopeptides **10** and **11** exhibited decreased activity in comparison with the native α-mating factor. Interestingly, removal of the sialic acid residues in the complex type glycopeptide **10** by sialidase treatment restored its biological activity. On the other hand, peptide **9** with a single GlcNAc exhibited 2-fold higher activity than the native factor. The data again suggest that distinct glycosylation of a given bioactive peptide can modulate its biological activity. As expected, the glycosylated α-mating factors were found to be more stable against proteolysis than the native α-mating factor.

Peptide T was a small peptide consisting of eight amino acid residues that was reportedly active in blocking the infection of human T cells by HIV virus (*41*). Yamamoto et al (*42*) demonstrated an efficient synthesis of a glycosylated peptide T in which a complex type oligosaccharide was added to the consensus N-glycosylation site (Figure 2). The yield for the Endo-M catalyzed transglycosylation is 39% when 2-fold of the glycosyl donor was used. The resulting glycosylated peptide T was highly stable against proteolysis in comparison with peptide T itself.

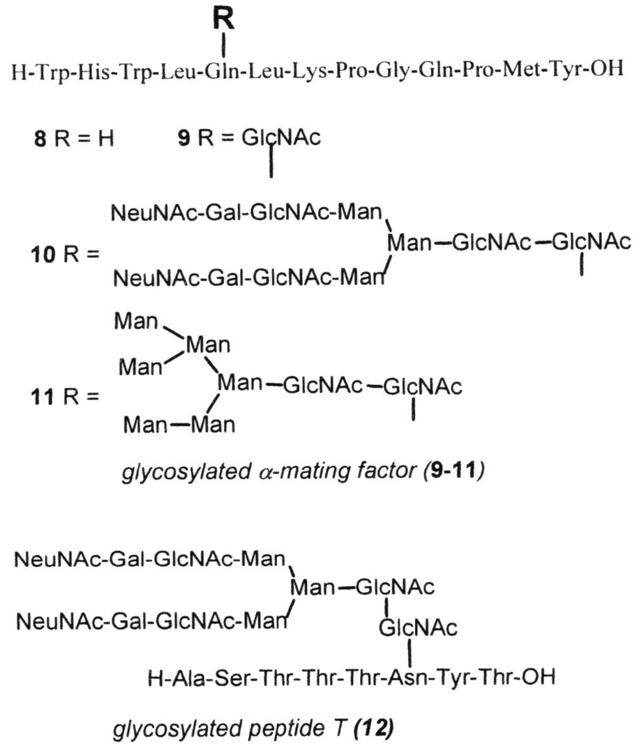

R
H-Trp-His-Trp-Leu-Gln-Leu-Lys-Pro-Gly-Gln-Pro-Met-Tyr-OH

8 R = H **9** R = GlcNAc

10 R = NeuNAc-Gal-GlcNAc-Man / NeuNAc-Gal-GlcNAc-Man — Man—GlcNAc—GlcNAc

11 R = Man, Man — Man; Man—Man — Man—GlcNAc—GlcNAc

*glycosylated α-mating factor (**9-11**)*

NeuNAc-Gal-GlcNAc-Man / NeuNAc-Gal-GlcNAc-Man — Man—GlcNAc — GlcNAc
H-Ala-Ser-Thr-Thr-Thr-Asn-Tyr-Thr-OH

*glycosylated peptide T (**12**)*

Figure 2. Structures of glycosylated α-mating factor and peptide T

Glycoforms of Complex Glycopeptides for Conformational Studies

It is known that N-glycosylation influences the protein folding process. Although some model glycopeptides with mono- or di-saccharide moieties attached are useful for probing the effects of glycosylation on peptide structure, the studies to establish the impact of native, large oligosaccharides on peptide and protein conformation have been limited due to the difficulties in obtaining homogeneous, complex glycopeptides of interest. Taking the power of Endo-M catalyzed transglycosylation, O'Connor et al (*43*) was able to add a complex type oligosaccharide chain on an extra-cellular loop peptide from the α-subunit of the nicotinic acetylcholine receptor (nAChR). Thus, using a complex type glycopeptide isolated from the pronase digestion of human transferrin as the glycosyl donor, and the GlcNAc-containing peptide **14** as the acceptor, the Endo-M catalyzed reaction gave the sialo complex type glycopeptide **16**. In addition, sequential treatment of **16** with sialidase and galactosidase to remove the outer sialic acid and galactose residues generated additional homogeneous glycopeptides **17** and **18**, respectively (Figure 3). With the availability of these homogeneous glycopeptides, O'Connor et al investigated the solution structures of the peptides using homonuclear 2D NMR techniques. The study suggests that addition of N-linked saccharide affects certain key steps of protein folding, such as disulfide bond formation and *cis/trans* proline isomerization. Comparison of the conformational effects of truncated saccharides **14** and **15** with large native oligosaccharide structures **16-18** revealed that the first two GlcNAc residues that are in immediate proximity to the peptide backbone mediate the bulk of the conformational effect. However, the addition of the outer saccharide residues also exhibited a number of spectroscopic characteristics that revealed a modest, detectable structural effect (*43*). Taken together, the data suggest that, in addition to the immediate inner sugar residues linked to the protein backbone, the outer sugar residues in the native *N*-oligosaccharides may also modulate the process of protein folding *in vivo*.

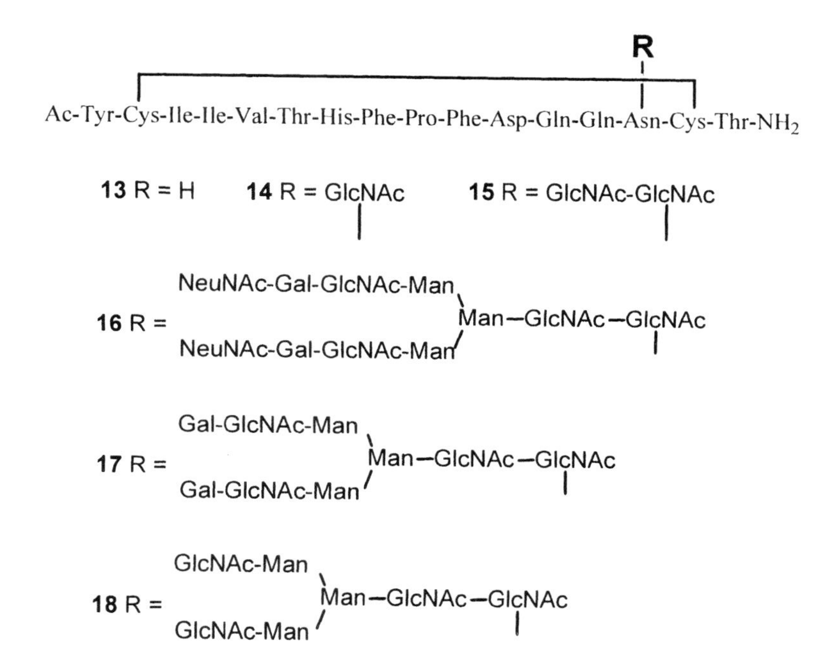

Figure 3. Glycoforms of an extracellular loop peptide from nAChR

Endoglycosidase-Catalyzed Synthesis of Substrate Analogs for Glycoamidases

Glycoamidase, also called peptide N-glycanase (PNGase) or glycopeptidase, is a class of enzymes that release the intact oligosaccharides from N-glycoproteins by cleaving the amide linkage between the carbohydrate and protein. The wide occurrence of glycoamidase activity in microorganism, plants, and animals suggests that the glycoamidase-mediated protein de-glycosylation may play a role in biological processes (*44*).

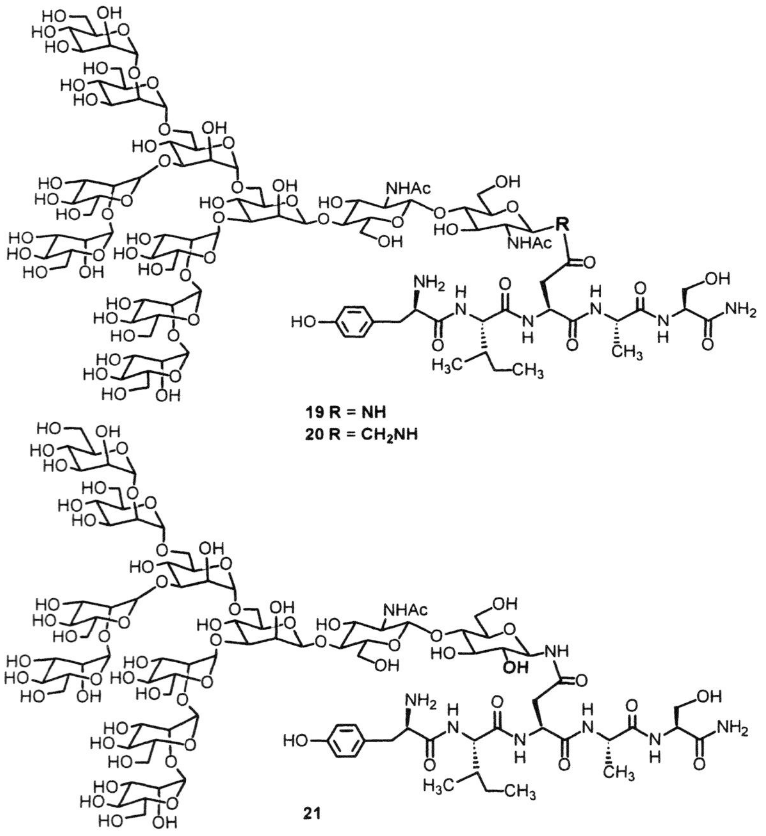

19 R = NH
20 R = CH_2NH

21

Figure 4. Structures of glycopeptide substrate analogs for glycoamidase

Glycoamidases recognize both the oligosaccharide and the peptide/protein portions for activity. To probe the mechanism of action and biological functions of glycoamidases, glycopeptides and analogs containing a large oligosaccharide chain are needed. Endoglycosidase-catalyzed transglycosylation has found to be particularly useful for the construction of the molecular probes. Wang et al (*45,46*) designed a novel C-glycopeptide as a substrate analog. The C-glycopeptide **20** contains all the structural features of the corresponding natural N-glycopeptide **19** except the insertion of a methylene group at the crucial asparagine-GlcNAc linkage (Figure 4). Deras et al (47) synthesized a novel glycopeptide **21**, where the N-linked GlcNAc was replaced with a *Glc* moiety (Figure 4). The chemoenzymatic synthesis of **20** was summarized in Scheme 3.

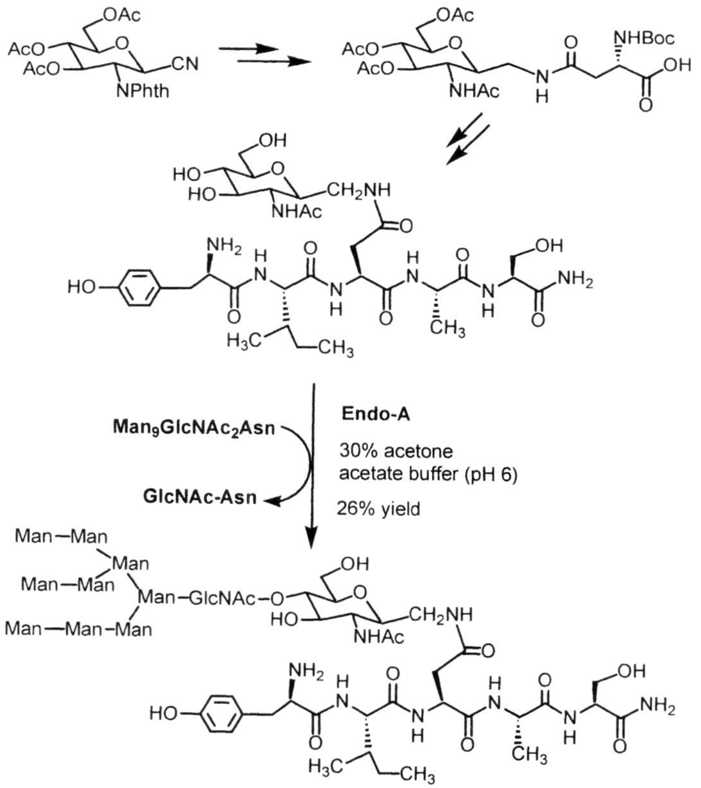

Scheme 3. Synthesis of the C-glycopeptide ***20****.*

To prepare the GlcNAc-C-peptide acceptor, a C-linked asparagine residue was first introduced at the anomeric position of the glucosamine moiety. Then the peptide chain was elongated from the N- and C-terminus and protecting groups were subsequently removed to give the GlcNAc-C-peptide. A high-mannose type oligosaccharide moiety was then transferred from $Man_9GlcNAc_2Asn$ (isolated from soybean agglutinin via pronase digestion) to the acceptor under the action of Endo-A to give the desired C-glycopeptide **20** (Scheme 3). Addition of acetone into the reaction buffer increased the efficiency of the enzymatic transglycosylation by depressing the competing hydrolysis. The C-glycopeptide was isolated in 26% yield (based on glycosyl donor) when 4-fold excess of the glycosyl acceptor was used. Excess acceptor could quantitatively recovered during HPLC purification and could be re-used. The Endo-A catalyzed synthesis of the C-glycopeptide was found to be equally efficient as for the synthesis of the natural N-glycopeptide **19**. This implicates that Endo-A does not discriminate the functional group (C- vs. N-linked) at the anomeric position of the GlcNAc moiety in the acceptor.

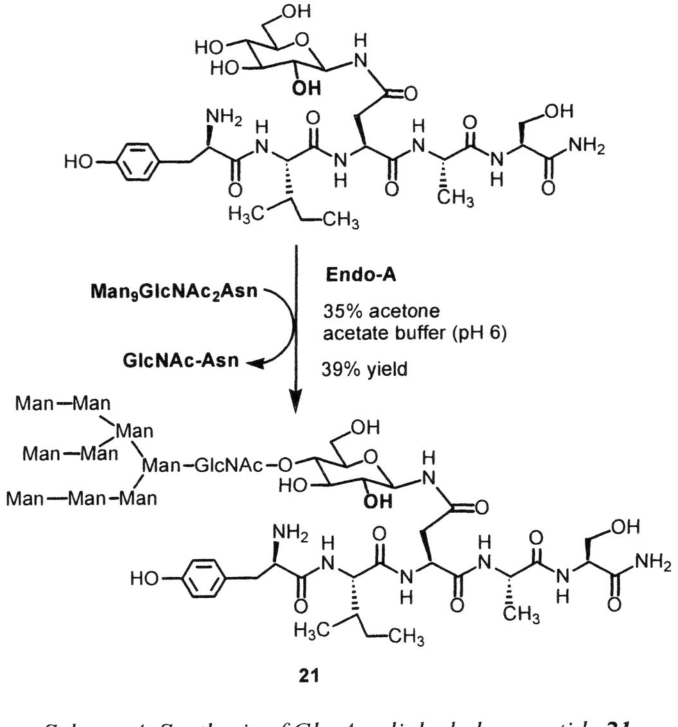

$Man_9GlcNAc_2Asn$ → GlcNAc-Asn

Endo-A
35% acetone
acetate buffer (pH 6)
39% yield

21

*Scheme 4. Synthesis of Glc-Asn-linked glycopeptide **21**.*

On the other hand, Deras et al (*47*) synthesized a high-mannose type glycopeptide analog containing a glucose-asparagine linkage (compound **21**, Figure 4). The synthesis takes advantage of the fact that Endo-A can take glucose as an acceptor in the transglycosylation reaction (Scheme 4).

The biological activities of the synthetic substrate analogs **20** and **21** towards various glycoamidases were studied. It was found that neither the C-linked **20** nor the Glc-Asn-linked **21** are substrates for glycoamidases. Instead, both analogs are competitive inhibitors against all glycoamidases tested (*46,47*).

Synthesis of HIV-1 gp120 Glycopeptides Related to the Epitope of the Neutralizing Antibody 2G12

The human immunodeficiency virus type 1 (HIV-1) envelope glycoprotein gp120 is heavily glycosylated. It has 24 conserved N-glycosylation sites, where 13 are complex-type N-glycans, and 11 are high-mannose type and/or hybrid type N-glycans. (*48-50*). Human monoclonal antibody 2G12 is one of the few antibodies so far identified that are capable of neutralizing a broad range of HIV-1 primary isolates. Moreover, 2G12 is the only antibody that recognizes a novel carbohydrate antigen on gp120 (*51*). Early mutational studies revealed that the high-mannose type oligosaccharides at the glycosylation sites of 295N, 332N, 339N, 386N, and 392N were involved in the binding to 2G12. More recent studies suggest that the epitope of 2G12 is likely to be a novel oligosaccharide cluster consisting of two or more high-mannose sugar chains at the N-glycosylation sites (*51-53*).

It was reported that glycosylation on HIV-1 gp120 is diverse and highly heterogeneous. In addition to various complex type N-glycans, the high-mannose type N-glycans on gp120 range from Man_5 and Man_6 to Man_9 (*48,54*). To exploit the novel carbohydrate antigen for HIV-1 vaccine design, we initiated a project aiming to duplicate all the possible antigenic structures through synthesis and to investigate their binding to 2G12. A short peptide from gp120 sequence 336-342 was chosen as a model peptide that contains the N339 glycosylation site. Homogeneous glycoforms of the gp120 peptide containing Man_9, Man_6, and Man_5 N-glycans, respectively, were synthesized using endo-β-*N*-acetylglucosaminidase from *Arthrobacter* (Endo-A) as the key enzyme (Figure 5). Using $Man_9GlcNAc_2Asn$ as the glycosyl donor and the chemically synthesized *N*-acetylglucosaminyl peptide as the glycosyl acceptor, the Endo-A catalyzed transglycosylation gave the Man_9-glycopeptide **24** in 28% isolated yield (*55*). Similarly, Transglycosylation using $Man_5GlcNAc_2Asn$ and $Man_6GlcNAc_2Asn$ as the glycosyl donors, which were prepared through pronase digestion of chicken ovalbumin, gave the Man_5-and Man_6-glycopeptides **22** and **23** in 25% and 27% yields, respectively (Figure 5). No difference was observed in the Endo-A activities toward the different high-mannose type glycosyl donors (*55*). Therefore, Endo-A is particularly useful for the construction of various high-mannose type glycoforms of glycopeptides.

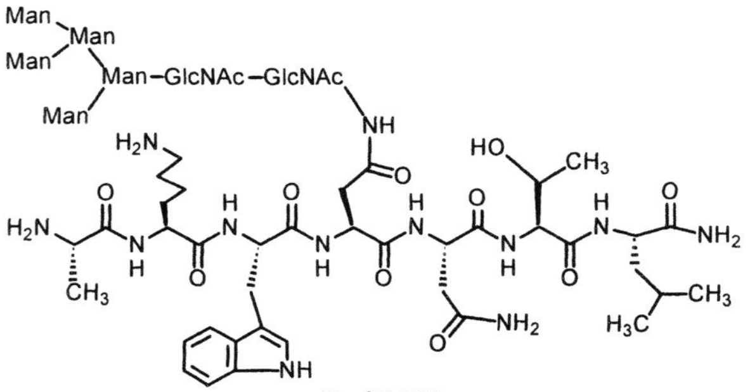

M_5-GP (22)

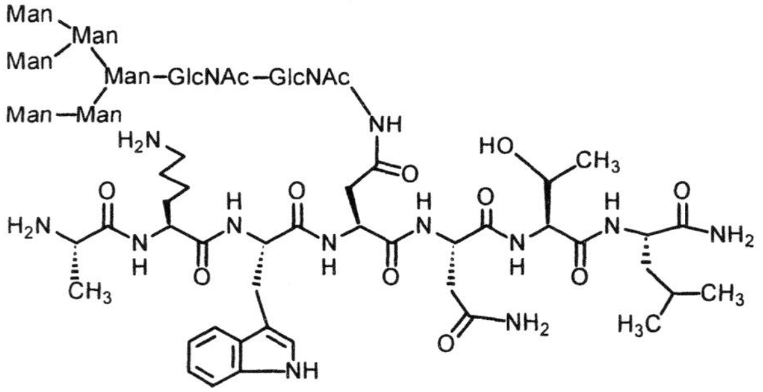

M_6-GP (23)

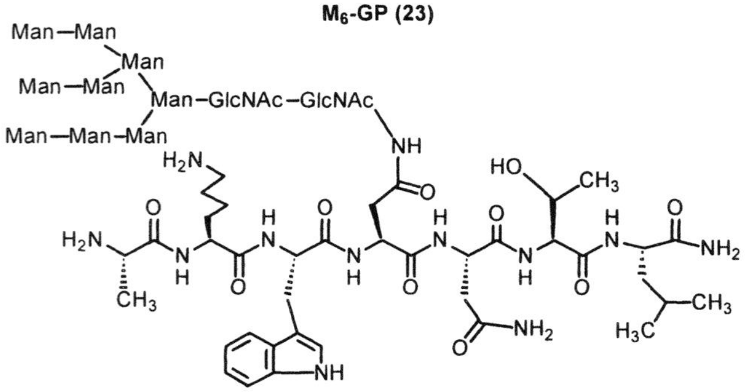

M_9-GP (24)

Figure 5. Structures of the synthetic high-mannose type gp120 glycopeptides.

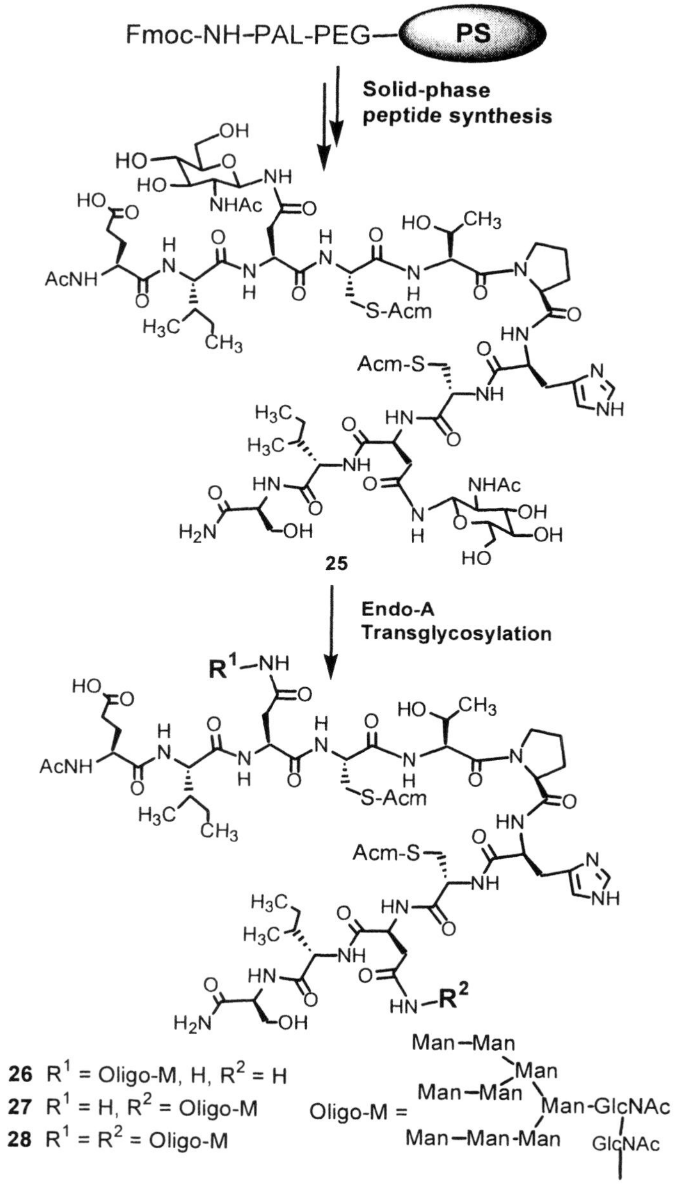

Fmoc-NH-PAL-PEG—PS

Solid-phase peptide synthesis

25

Endo-A Transglycosylation

26 R^1 = Oligo-M, H, R^2 = H

27 R^1 = H, R^2 = Oligo-M

28 R^1 = R^2 = Oligo-M

Oligo-M = Man–Man–Man, Man–Man–Man, Man–Man–Man–Man-GlcNAc-GlcNAc

Scheme 5. Synthesis of double-glycosylated gp120 glycopeptide

Next, we attempted to synthesize a large glycopeptide from the gp120 sequence 293-334 that contains two N-glycans at the glycosylation sites N295 and N332, which are most likely to form the actual epitope for 2G12 (52). To simplify the synthesis, we replaced the 32 amino acid residues in the V3 loop with a proline residue, because deletion of the V3 loop in native gp120 did not affect the binding of 2G12 to gp120 (51-53). The synthesis of the double-glycosylated gp120 glycopeptide **28** was summarized in Scheme 5. The undecapeptide **25** containing two GlcNAc moieties at the putative N295 and N332 sites was synthesized by solid phase peptide synthesis and purified by HPLC [ESI-MS of **25**: 1820.90, $(M+H)^{+}$; 910.74, $(M+2H)^{2+}$; 607.58, $(M+3H)^{3+}$] (Wang et al, unpublished results). Using $Man_9GlcNAc_2Asn$ as the glycosyl donor, the Endo-A catalyzed transfer of a Man_9 moiety to the acceptor **25** was monitored by HPLC. Three transglycosylation products were formed, which were purified by HPLC and characterized by ESI-MS. The major products isolated are the mono-glycosylated glycopeptides **26** and **27**, which are indistinguishable on HPLC and isolated in 32% yield [ESI-MS of **26** and **27**: 1742.41, $(M+2H)^{2+}$; 1161.82, $(M+3H)^{3+}$]. The desired, double-glycosylated glycopeptide **28** was isolated in 2.5% yield [ESI-MS of **28**: 1715.76, $(M+3H)^{3+}$; 1287.06, $(M+4H)^{4+}$] (Wang et al, unpublished results). The low yield of **28** may be partially due to the steric hindering for the addition of the second oligosaccharide moiety. Folding of the synthetic glycopeptides and the study on their binding to antibody 2G12 are underway. To the best of our knowledge, this is perhaps the first example of double-glycosylation of a large peptide through the endoglycosidase-catalyzed reaction.

Conclusion

The advantage of using the endoglycosidase-catalyzed reaction for glycopeptide synthesis is apparent. In contrast to glycosyltransferase, which adds monosaccharide units one at a time during a synthesis, the endoglycosidase transfers an oligosaccharide moiety to an acceptor in a single step. The transglycosylation ability of the endo-β-*N*-acetylglucosamindases has made it possible to construct large and interesting glycopeptides for structural and biological studies. The endoglycosidase-catalyzed transglycosylation was also useful for glycosylation remodeling of native glycoproteins, as demonstrated by Takegawa et al for the remodeling of ribonuclease B (*56,57*). Endoglycosidase-catalyzed synthesis requires a pre-attachment of a monosaccharide moiety (GlcNAc) at the desired glycosylation site in the peptide or protein. However, GlcNAc-peptides can be efficiently prepared by the established solid phase peptide synthesis using GlcNAc-Asn or other monosaccharide-amino acid as building blocks without special difficulties. Even very large GlcNAc-containing peptides or proteins can be synthesized when native chemical ligation (*58*) or protease-catalyzed ligation (*59*) is incorporated in the synthetic scheme. In addition, site-specific incorporation of monosaccharides into proteins is also achievable through molecular biology, using the nonsense codon suppression read-through techniques (*60*) with artificial glyco-aminoacyl-tRNAs (*61-64*). It should be pointed out that endoglycosidases inherently are hydrolytic enzymes. Therefore, a major drawback in the endoglycosidase-catalyzed reaction is the relatively low yield, because the resulting product also becomes a substrate for the endoglycanase-catalyzed hydrolysis. Enzyme/protein engineering may offer a hope in improving the efficiency, as exemplified in the attempted generation of glyco-synthases from glycosidases (*65,66*). Another limitation seems to come from the glycosyl donors. The endoglycosidase-catalyzed syntheses have been so far limited to the use of oligosaccharides or glycopeptides isolated from natural source as glycosyl donors. However, we predict with confidence that future exploitation of various synthetic substrates (*67*) as glycosyl donors for endoglycosidases, in combination with enzyme/protein engineering, will reinforce the usefulness of the synthetic strategy. With continuous efforts in the direction, the endoglycosidase-catalyzed transglycosylation will certainly become a powerful alternative for the synthesis and remodeling of complex glycopeptides and glycoproteins.

Acknowledgment

We thank Professor Kaoru Takegawa (Kagawa University) for kindly providing us the pGEX-2T/Endo-A plasmid for overproducing the endo-enzyme. Our work on chemoenzymatic synthesis was financially supported by the Institute of Human Virology, University of Maryland Biotechnology Institute.

References

1. Varki, A. *Glycobiology* **1993**, *3*, 97-130.
2. Dwek, R. A. *Chem. Rev.* **1996**, *96*, 683-720.
3. Rudd, P. M.; Elliott, T.; Cresswell, P.; Wilson, I. A.; Dwek, R. A. *Science* **2001**, *291*, 2370-2376.
4. Arsequell, G.; Valencia, G. *Tetrahedron: Asymmetry* **1999**, *10*, 3045-3094.
5. Seitz, O. *CHEMBIOCHEM* **2000**, *1*, 214-246.
6. Davis, B. G. *Chem. Rev.* **2002**, *102*, 579-601.
7. Sears, P.; Tolbert, T.; Wong, C. H. *Genet. Eng.* **2001**, *23*, 45-68.
8. Roberge, J. Y.; Beebe, X.; Danishefsky, S. J. *Science* **1995**, *269*, 202-204.
9. Wong, C. H. *Acta. Chem. Scand.* **1996**, *50*, 211-218.

10. Watt, G. M.; Lowden, P. A.; Flitsch, S. L. *Curr. Opin. Struct. Biol.* **1997**, *7*, 652-660.
11. Crout, D. H. G.; Vic, G. *Curr. Opin. Chem. Biol.* **1998**, *2*, 98-111.
12. Wymer, N.; Toone, E. J. *Curr. Opin. Chem. Biol.* **2000**, *4*, 110-119.
13. Li, Y. T.; Carter, B. Z.; Rao, B. N.; Schweingruber, H.; Li, S. C. *J. Biol. Chem.* **1991**, *266*, 10723-10726.
14. Yamada, K.; Fujita, E.; Nishimura, S. *Carbohydr. Res.* **1997**, *305*, 443-461.
15. Kobayashi, S.; Kiyosada, T.; Shoda, S. *J. Am. Chem. Soc.* **1996**, *118*, 13113-13114.
16. Moreau, V.; Driguez, H. *J. Chem. Soc. Perkin Trans I* **1996**, 525-527.
17. Kobayashi, S.; Morii, H.; Itoh, R.; Kimura, S.; Ohmae, M. *J. Am. Chem. Soc.* **2001**, *123*, 11825-11826.
18. Ajisaka, K.; Miyasato, M.; Ishii-Karakasa, I. *Biosci. Biotechnol. Biochem.* **2001**, *65*, 1240-1243.
19. Yamamoto, K. *J. Biosci. Bioeng.* **2001**, *92*, 493-501.
20. Trimble, R. B.; Atkinson, P. H.; Tarentino, A. L.; Plummer, T. H., Jr.; Maley, F.; Tomer, K. B. *J. Biol. Chem.* **1986**, *261*, 12000-12005.
21. Takegawa, K.; Yamaguchi, S.; Kondo, A.; Iwamoto, H.; Nakoshi, M.; Kato, I.; Iwahara, S. *Biochem. Int.* **1991**, *24*, 849-855.
22. Takegawa, K.; Yamaguchi, S.; Kondo, A.; Kato, I.; Iwahara, S. *Biochem. Int.* **1991**, *25*, 829-835.
23. Takegawa, K.; Fujita, K.; Fan, J. Q.; Tabuchi, M.; Tanaka, N.; Kondo, A.; Iwamoto, H.; Kato, I.; Lee, Y. C.; Iwahara, S. *Anal. Biochem.* **1998**, *257*, 218-223.
24. Kadowaki, S.; Yamamoto, K.; Fujisaki, M.; Izumi, K.; Tochikura, T.; Yokoyama, T. *Agric. Biol. Chem.* **1990**, *54*, 97-106.
25. Kadowaki, S.; Yamamoto, K.; Fujisaki, M.; Tochikura, T. *J. Biochem. (Tokyo)* **1991**, *110*, 17-21.
26. Yamamoto, K.; Kadowaki, S.; Fujisaki, M.; Kumagai, H.; Tochikura, T. *Biosci. Biotechnol. Biochem.* **1994**, *58*, 72-77.
27. Yamamoto, K.; Kadowaki, S.; Watanabe, J.; Kumagai, H. *Biochem. Biophys. Res. Commun.* **1994**, *203*, 244-252.
28. Haneda, K.; Inazu, T.; Yamamoto, K.; Kumagai, H.; Nakahara, Y.; Kobata, A. *Carbohydr. Res.* **1996**, *292*, 61-70.
29. Fan, J. Q.; Takegawa, K.; Iwahara, S.; Kondo, A.; Kato, I.; Abeygunawardana, C.; Lee, Y. C. *J. Biol. Chem.* **1995**, *270*, 17723-17729.
30. Fan, J. Q.; Quesenberry, M. S.; Takegawa, K.; Iwahara, S.; Kondo, A.; Kato, I.; Lee, Y. C. *J. Biol. Chem.* **1995**, *270*, 17730-17735.
31. Nicholson, G. C.; Moseley, J. M.; Sexton, P. M.; Mendelsohn, F. A.; Martin, T. J. *J. Clin. Invest.* **1986**, *78*, 355-360.
32. Ogawa, K.; Nishimura, S.; Doi, M.; Kyogoku, Y.; Hayashi, M.; Kobayashi, Y. *Eur. J. Biochem.* **1994**, *222*, 659-666.

33. Haneda, K.; Inazu, T.; Mizuno, M.; Iguchi, R.; Yamamoto, K.; Kumagai, H.; Aimoto, S.; Suzuki, H.; Noda, T. *Bioorg. Med. Chem. Lett.* **1998**, *8*, 1303-1306.
34. Mizuno, M.; Haneda, K.; Iguchi, R.; Muramoto, I.; Kawakami, T.; Aimoto, S.; Yamamoto, K.; Inazu, T. *J. Am. Chem. Soc.* **1999**, *121*, 284-290.
35. Yamamoto, K.; Haneda, K.; Iguchi, R.; Inazu, T.; Mizuno, M.; Takegawa, K.; Kondo, A.; Kato, I. *J. Biosci. Bioeng.* **1999**, *87*, 175-179.
36. Hashimoto, Y.; Toma, K.; Nishikido, J.; Yamamoto, K.; Haneda, K.; Inazu, T.; Valentine, K. G.; Opella, S. J. *Biochemistry* **1999**, *38*, 8377-8384.
37. Tagashira, M.; Tanaka, A.; Hisatani, K.; Isogai, Y.; Hori, M.; Takamatsu, S.; Fujibayashi, Y.; Yamamoto, K.; Haneda, K.; Inazu, T.; Toma, K. *Glycoconj. J.* **2001**, *18*, 449-455.
38. Marx, J. L. *Science* **1979**, *205*, 886-889.
39. Haneda, K.; Inazu, T.; Mizuno, M.; Iguchi, R.; Tanabe, H.; Fujimori, K.; Yamamoto, K.; Kumagai, H.; Tsumori, K.; Munekata, E. *Biochim. Biophys. Acta* **2001**, *1526*, 242-248.
40. Saskiawan, I.; Mizuno, M.; Inazu, T.; Haneda, K.; Harashima, S.; Kumagai, H.; Yamamoto, K. *Arch. Biochem. Biophys.* **2002**, *406*, 127-134.
41. Pert, C. B.; Hill, J. M.; Ruff, M. R.; Berman, R. M.; Robey, W. G.; Arthur, L. O.; Ruscetti, F. W.; Farrar, W. L. *Proc. Natl. Acad. Sci. USA* **1986**, *83*, 9254-9258.
42. Yamamoto, K.; Fujimori, K.; Haneda, K.; Mizuno, M.; Inazu, T.; Kumagai, H. *Carbohydr. Res.* **1998**, *305*, 415-422.
43. O'Connor, S. E.; Pohlmann, J.; Imperiali, B.; Saskiawan, I.; Yamamoto, K. *J. Am. Chem. Soc.* **2001**, *123*, 6187-6188.
44. Suzuki, T.; Kitajima, K.; Inoue, S.; Inoue, Y. In *Glycosciences: Status and Perspectives*; Gabius, H. J., Gabius, S., Eds.; Chapman & Hall GmbH: Weinheim, Germany, 1997; pp 121-131.
45. Wang, L. X.; Fan, J. Q.; Lee, Y. C. *Tetrahedron Lett.* **1996**, *37*, 1975-1978.
46. Wang, L. X.; Tang, M.; Suzuki, T.; Kitajima, K.; Inoue, Y.; Inoue, S.; Fan, J. Q.; Lee, Y. C. *J. Am. Chem. Soc.* **1997**, *119*, 11137-11146.
47. Deras, I. L.; Takegawa, K.; Kondo, A.; Kato, I.; Lee, Y. C. *Bioorg. Med. Chem. Lett.* **1998**, *8*, 1763-1766.
48. Mizuochi, T.; Matthews, T. J.; Kato, M.; Hamako, J.; Titani, K.; Solomon, J.; Feizi, T. *J. Biol. Chem.* **1990**, *265*, 8519-8524.
49. Leonard, C. K.; Spellman, M. W.; Riddle, L.; Harris, R. J.; Thomas, J. N.; Gregory, T. J. *J. Biol. Chem.* **1990**, *265*, 10373-10382.
50. Zhu, X.; Borchers, C.; Bienstock, R. J.; Tomer, K. B. *Biochemistry* **2000**, *39*, 11194-11204.
51. Trkola, A.; Purtscher, M.; Muster, T.; Ballaun, C.; Buchacher, A.; Sullivan, N.; Srinivasan, K.; Sodroski, J.; Moore, J. P.; Katinger, H. *J. Virol.* **1996**, *70*, 1100-1108.

52. Scanlan, C. N.; Pantophlet, R.; Wormald, M. R.; Ollmann Saphire, E.; Stanfield, R.; Wilson, I. A.; Katinger, H.; Dwek, R. A.; Rudd, P. M.; Burton, D. R. *J. Virol.* **2002**, *76*, 7306-7321.
53. Sanders, R. W.; Venturi, M.; Schiffner, L.; Kalyanaraman, R.; Katinger, H.; Lloyd, K. O.; Kwong, P. D.; Moore, J. P. *J. Virol.* **2002**, *76*, 7293-7305.
54. Geyer, H.; Holschbach, C.; Hunsmann, G.; Schneider, J. *J. Biol. Chem.* **1988**, *263*, 11760-11767.
55. Singh, S.; Ni, J.; Wang, L. X. *Bioorg. Med. Chem. Lett.* **2003**, *13*, 327-330.
56. Takegawa, K.; Tabuchi, M.; Yamaguchi, S.; Kondo, A.; Kato, I.; Iwahara, S. *J. Biol. Chem.* **1995**, *270*, 3094-3099.
57. Fujita, K.; Tanaka, N.; Sano, M.; Kato, I.; Asada, Y.; Takegawa, K. *Biochem. Biophys. Res. Commun.* **2000**, *267*, 134-138.
58. Hackeng, T. M.; Griffin, J. H.; Dawson, P. E. *Proc. Natl. Acad. Sci. USA* **1999**, *96*, 10068-10073.
59. Witte, K.; Sears, P.; Martin, R.; Wong, C. H. *J. Am. Chem. Soc.* **1997**, *119*, 2114-2118.
60. Noren, C. J.; Anthony-Cahill, S. J.; Griffith, M. C.; Schultz, P. G. *Science* **1989**, *244*, 182-188.
61. Mamaev, S. V.; Laikther, A. L.; Arslan, T.; Hecht, S. M. *J. Am. Chem. Soc.* **1996**, *118*, 7243-7244.
62. Arslan, T.; Mamaev, S. V.; Mamaeva, N. V.; Hecht, S. M. *J. Am. Chem. Soc.* **1997**, *119*, 10877-10887.
63. Schmidt, R. R.; Castro-Palomino, J. C.; Retz, O. *Pure Appl. Chem.* **1999**, *71*, 729-735.
64. Fahmi, N. E.; Golovine, S.; Wang, B.; Hecht, S. M. *Carbohydr. Res.* **2001**, *330*, 149-164.
65. Mackenzie, L. F.; Wang, Q. P.; Warren, R. A. J.; Withers, S. G. *J. Am. Chem. Soc.* **1998**, *120*, 5583-5584.
66. Tolborg, J. F.; Petersen, L.; Jensen, K. J.; Mayer, C.; Jakeman, D. L.; Warren, R. A.; Withers, S. G. *J. Org. Chem.* **2002**, *67*, 4143-4149.
67. Fujita, M.; Shoda, S.; Haneda, K.; Inazu, T.; Takegawa, K.; Yamamoto, K. *Biochim. Biophys. Acta* **2001**, *1528*, 9-14.

Chapter 7

Strategies for Synthesis of an Oligosaccharide Library Using a Chemoenzymatic Approach

Ola Blixt and Nahid Razi

Carbohydrate Synthesis and Protein Expression Core Resource, Consortium for Functional Glycomics, The Scripps Research Institute, Department of Molecular Biology CB216, 10550 North Torrey Pines Road, La Jolla, CA 92037

The expanding interest for carbohydrates and glycoconjugates in cell communication has led to an increased demand of these structures for biological studies. Complicated chemical strategies in carbohydrate synthesis are now more frequently replaced by regiospecific and stereospecific enzymes. The exploration of microbial resources and improved production of mammalian enzymes have introduced glycosyltransferases as an efficient complementary tool in carbohydrate synthesis. In this report we will discuss preparative synthesis of different categories of glycans, such as fucosylated and/or sialylated derivatives of galactosides/polylactosamines and ganglioside mimic structures. We also introduce a feasible procedure to incorporate sialic acid derivatives such as *N*-glycolylneuraminic acid (Neu5Gc) and 3-deoxy-D-*glycero*-D-*galacto*-2-nonulosonic acid (KDN) into different *N*- and *O*-linked oligosaccharides.

Carbohydrate groups of glycoproteins and glycolipids exhibit tremendous structural diversity and are major components of the outer surface of animal cells(1). The lack of availability and affordability of appropriate oligosaccharides has always been a major limitation in carbohydrate research. Our goal in this study is to introduce an efficient and rapid production of oligosaccharides in sufficient amounts to build up a compound library. Such libraries will significantly contribute to more comprehensive studies involving carbohydrates and carbohydrate-binding proteins (CBPs) and, thus, advance the research in the field of glycobiology.

Chemoenzymatic Synthesis Strategy

Sugars are multifunctional compounds with several hydroxyl (-OH) groups of equal chemical reactivity. The manipulation of a single selected hydroxyl group is often difficult. Blocking one hydroxyl group and leaving one free is not trivial and requires a careful design of reactions to obtain the desired regioselectivity and stereoselectivity. The number of steps increase with the size of the oligosaccharide so that synthesis of a disaccharide may require 5 to 12 steps, and as many as 40 chemical steps can be involved in a typical tetrasaccharide synthesis, resulting in purification problems, low yields and high costs. Despite recent promising advances in chemical carbohydrate synthesis (2, 3), chemical synthesis of oligosaccharides is still very time-consuming and requires special expertise (4, 5).

One way to avoid the protection-deprotection steps entirely is to mimic nature's way of synthesizing oligosaccharides by using regiospecific and stereospecific enzymes, glycosyltransferases, for coupling reactions between the monosaccharides. These enzymes catalyze the transfer of a monosaccharide from a glycosyl donor (usually a sugar nucleotide) to a glycosyl acceptor with high efficiency. Most enzymes operate at room temperature in aqueous solutions (pH 6-8), which makes it possible to combine several enzymes in one pot for multi-step reactions. The high regioselectivity, stereoselectivity and catalytic efficiency make enzymes especially useful for practical synthesis of oligosaccharides and glycoconjugates (6-8).

The major setback for a breakthrough using of glycosyltransferases in synthetic applications have predominantly been the lack of available enzymes in sufficient amounts. Additionally, glycosyltransferases are generally very specific for the sugar nucleotide and acceptor substrates and therefore most of the synthetic applications of glycosyltransferases are centered on the assembly of natural occurring oligosaccharides found in mammalian systems. Most of the enzymes are prepared from mammalian tissues by isolation or by gene cloning and expression in various expression systems such as mammalian, insect cell, yeast or bacteria (9, 10). Recent advances in isolating and cloning glycosyltransferases from non-mammalian sources such as bacteria have greatly enhanced the possibility for an increased production of various oligosaccharides (11-13). Bacterial glycosyltransferases have turned out to be more relaxed regarding donor and acceptor specificities. Furthermore, these enzymes are well expressed in bacterial expression systems such as *E. coli* that can easily be scaled up for over expression of the enzymes. Bacterial expression systems lack the post-translational modification machinery that is required for correct folding and activity of the mammalian enzymes whereas the enzymes from the bacterial sources are compatible with this system. This makes the bacterial enzymes more favorable as synthetic tools in glycobiology than the enzymes from the mammalian sources.

The following section will cover several synthetic approaches used to produce different categories of carbohydrates with a tool box of recombinant glycosyltransferases expressed in our laboratory.

Generating the Carbohydrate Compound Library

Our approach has been to use both chemical and, to as large extent as possible, enzymatic routes to synthesize a diverse compound library. Chemical steps were limited to the preparation of key core structures such as lactose- (Galβ(1-4)Glc-), N-acetylgluosamine- (GlcNAc-), iso-N-acetyllactosamine- (Galβ(1-3)GlcNAc-), T- (Galβ3GalNAcα-) and Tn- (GalNAcα-) antigen structures in multigram scale quantities (1-50 g). Most of these core structures are linked to a short neutral flexible spacer (sp_1 = $OCH_2CH_2N_3$) that enables a broad diversification such as attachment to proteins, solid supports (affinity matrices, polystyrene), biotin or other functional groups (Figure 1). The core units were later subject to enzymatic elongation using recombinant glycosyltransferases.

Enzymatic synthesis of carbohydrates in mg amounts is generally a straightforward procedure in terms of isolation and purification of the final structure. Ion exchange and size exclusion chromatography are the common procedures for product isolation. However, serious purification problems can be encountered when scaling up to multi gram reactions. Additives, such as nucleotide sugars, buffer salts, and crude enzymes are among the factors that interfere with efficient purification with the conventional chromatographic techniques used on mg-scale. However, different options are available to overcome these obstacles. One approach is to attach a hydrophobic reversible functional group at the reducing end and isolate the product by solid-phase extractions. This method is very efficient with fully completed reactions since the product will be the only isolated compound. However, in many cases

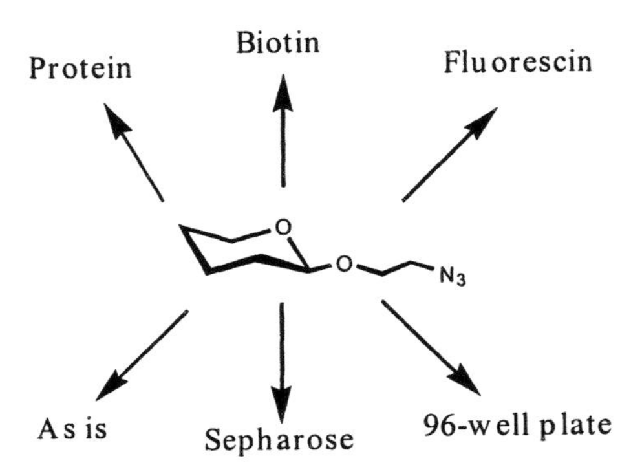

Figure 1. *Conjugations to various supports via the azido-functional group*

enzymatic reactions do not go to completion and the separation of the remaining starting material from product could be a problem. Another approach is to treat a completed enzymatic reaction mixture with acetic anhydride in pyridine to protect all free hydroxyl groups with acetyl groups including starting material, byproducts, and product. This crude mixture could then be partially purified with solvent extractions. Proteins, charged compounds, and salts will remain in the water phase, whereas the acetylated product will stay in the organic phase. For complete purification, the extracted mixture is purified with silica gel column chromatography. In general, the isolation strategy used depends on the scale and the compounds to be prepared.

Synthesis of Galactosides and Poly-*N*-acetyllactosamines

Numerous chemical approaches have been developed to prepare *N*-acetyllactosamines by glycosylation between derivatives of galactose and *N*-acetylglucosamine (14-16). Considering the tedious multiple protection/deprotection steps involved in chemical synthesis, the amounts of products obtained in these methods seldom exceed gram quantities. Enzymatically, the repeating Galβ(1-4)GlcNAc- unit is synthesized by the concerted action of β4-galactosyltransferase (β4GalT) and β3-*N*-acetyllactosamninyltransferase (β3GlcNAcT) (17, 18). We have previously cloned and characterized the bacterial *N. memingitidis* enzymes β4GalT-GalE (19) and β3GlcNAcT (20) and demonstrated their utility in preparative synthesis of various galactosides. β4GalT-GalE is a fusion protein constructed from β4GalT and the uridine-5'-diphospho-galactose-4'-epimerase (GalE) for *in situ* conversion of inexpensive UDP-glucose to UDP-galactose providing a cost efficient strategy. Both enzymes, β4GalT-GalE and β3GlcNAcT, were over expressed in *E. coli* AD202 in a large-scale fermentor (100 L). Bacteria were cultured in 2YT medium and induced with *iso*-propyl-thiogalactopyranoside (IPTG) to ultimately produce 8-10 g of bacterial cell paste / L cell media. The enzymes were then released from the cells by a microfluidizer and were solubilized in Tris buffer (25 mM, pH 7.5) containing manganese chloride (10 mM) and Triton X (0.25%) to reach enzymatic activities of about 50 U/L and 115 U/L of cell culture β4GalT-GalE and β3GlcNAcT, respectively.

The specificity studies of the β3GlcNAcT (Table I) reveal that lactose (**4**) is the better acceptor substrate (100%) while the enzyme shows just about 7-8% activity with N-acetyllactosamine (**6**). Nevertheless, adding the hydrophobic para-nitrophenyl ring as an aglycon to the reducing end of the acceptors will enhance the activity of the enzyme up to 10 fold (compare **4** with **5** and **6** with **7**). The increase in the enzyme activity by adding a hydrophobic aglycon to the acceptor sugar, though to the lesser extent, has also been shown for β4GalT (compare **12** with **13, 14**). The relaxed substrate specificity of these enzymes makes them very useful for preparative synthesis of various carbohydrate structures, including poly-*N*-acetyllactosamines.

Poly-*N*-acetyllactosamine is a unique carbohydrate structure composed of *N*-acetyllactosamine repeats that provides the backbone structure for additional modifications, such as sialylation and/or fucosylation. These extended oligosaccharides have been shown to be involved in various biological functions (21, 22) by interacting as a specific ligand to selectins (23, 24) or galectins (25, 26). Based on the specificity data in table I, we designed enzymatic synthesis of galactosides and polylactosamines in multi gram quantities, that was further modified using various fucosyltransferases (FUTs) (see figure 2).

Consistent with the procedure, we initiated a systematic gram-scale synthesis of different fucosylated lactosamine derivatives using the following recombinant fucosyltransferases, FUT-II, FUT-III, FUT-IV, FUT-V, and FUT-VI. All the above fucosyltransferases, except for FUT-V (expressed in *A. niger* (27), have been produced in the insect cell expression system and have been either partially purified on a GDP-sepharose affinity column or concentrated in a Tangential Flow Filtrator (TFF-MWCO 10k) as a crude enzyme mixture. Several FUTs have been thoroughly characterized in terms of substrate specificities and biological functions in different laboratories (27-32). The available specificity data in combination with our large scale production of recombinant FUTs made it possible to synthesize various precious fucosides in multigram quantities. Figure 2 demonstrates the general procedure to elongate the poly-LacNAc backbone and selected fucosylated structures using different FUTs and GDP-fucose. The yields for different stages of production were 75-90% for LeX (2 enzymatic steps), 45-50% for dimeric LacNAc structures (4 enzymatic steps) and 30-35% for trimeric lacNAc structures (6 enzymatic steps). Details of these syntheses will be published elsewhere.

Table I. Selected β4GalT-GalE and β3GlcNAcT Specificity Data

Acceptor	*Relative enzyme activity (%)*
	β(1-3)GlcNAcT-activity[#]
1 Gal	5
2 Galα-OpNP	102
3 Galβ-OpNP	16
4 Galβ(1-4)Glc	**100**
5 Galβ(1-4)Glcβ-OpNP	945
6 Galβ(1-4)GlcNAc	7
7 Galβ(1-4)GlcNAcβ-OpNP	74
8 Galβ(1-3)GlcNAc	5
	β(1-4)GalT-GalE-activity*
9 Glc	80
10 Glcβ-OpNP	60
11 $GlcNH_2$	30
12 GlcNAc	**100**
13 GlcNAcβ-OpNP	120
14 GlcNAcβ-Osp_1	360
15 GlcNAllocβ-sp_2	550

Abbreviations: pNP, para-nitrophenyl; sp_1, 2-azidoethyl; sp_2, 5-azido-3-oxapentyl, Alloc, allyloxycarbonyl

SOURCE: # Data selected from reference (20) * Data selected from reference (19)

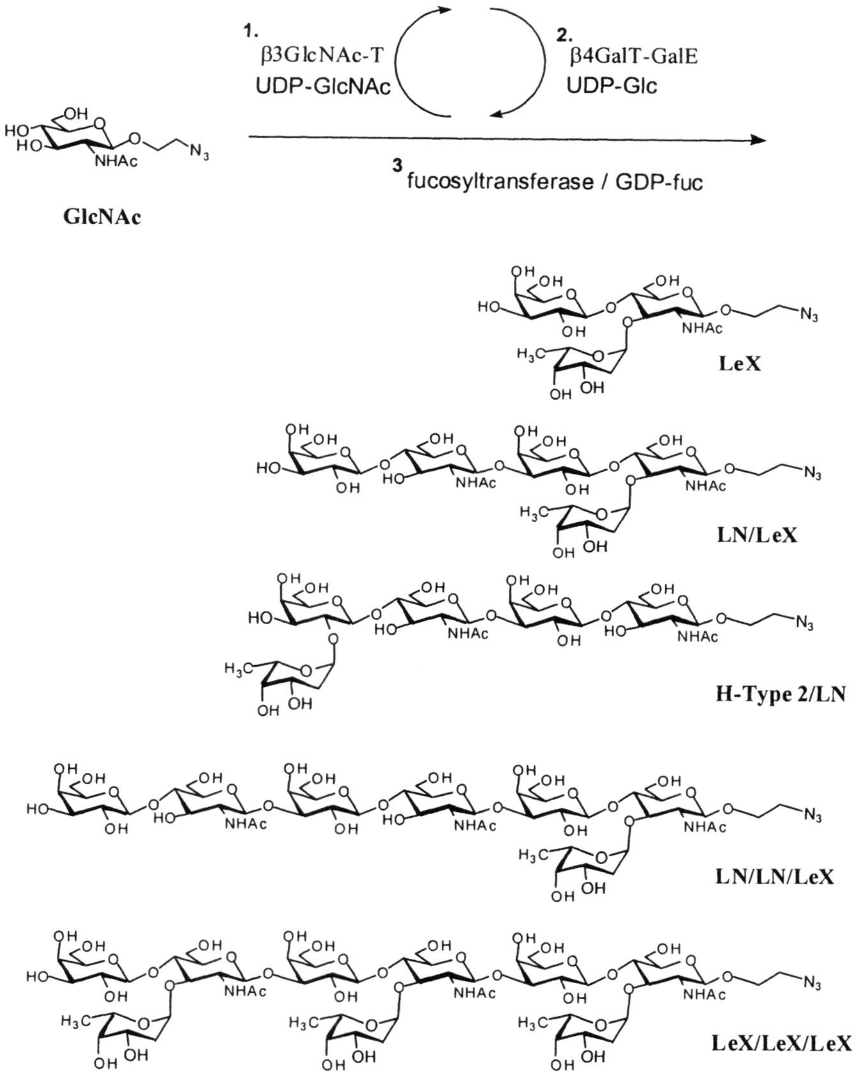

1. β3GlcNAc-T UDP-GlcNAc
2. β4GalT-GalE UDP-Glc
3 fucosyltransferase / GDP-fuc

GlcNAc

LeX

LN/LeX

H-Type 2/LN

LN/LN/LeX

LeX/LeX/LeX

Figure 2. *Selected syntheses of fucosylated N-acetyllactosamine structures*

Synthesis of sialic-acid-containing oligosaccharides

Sialic acid is a generic designation used for 2-keto-3-deoxy-nonulosonic acids. The most commonly occurring derivatives of this series of monosaccharides are those derived from *N*-acetylneuraminic acid (Neu5Ac), *N*-glycolylneuraminic acid (Neu5Gc) and the non-aminated 3-deoxy-D-*glycero*-D-*galacto*-2-nonulosonic acid (KDN). Sialic-acid-containing oligosaccharides are an important category of carbohydrates that are involved in different biological regulations and functions (33, 34). Sialic acids are shown to be involved in adsorption of toxins/viruses, and diverse cellular communications through interactions with carbohydrate binding proteins (CBPs) (35). Selectins and Siglecs (sialic acid-binding immunoglobulin-superfamily lectins) are among those well-characterized CBPs that function biologically through sialic acid interactions (36-38).

Synthesis of oligosaccharides containing sialic acids is not trivial. Unfortunately, the chemical approaches have several hampering factors in common. For example, stereo selective glycosylation with sialic acid generally gives an isomeric product, and as a result, purification problems and lower yields. Its complicated nature, also require extensive protecting group manipulations and careful design of both acceptor and donor substrates and substantial amounts of efforts are needed to prepare these building blocks (39-41).

For a fast and efficient way to sialylate carbohydrate structures, the method of choice is through catalysis by sialyltransferases. Enzymatic sialylation generating Neu5Ac-containing oligosaccharides have been a well-established route both in analytical and preparative scales (6, 42, 43). However, efficient methods for preparation of oligosaccharides having the Neu5Gc or KDN structures have not been explored to the same extent because of the scarcity of these sialoside derivatives.

With a modification of a method, originally developed by Wong and co-workers (38), we have established a simple way to obtain different sialoside derivatives, using recombinant sialyltransferases along with a commercial Neu5Ac aldolase. The preferred route to generate Neu5Ac-oligosaccharides was to use a one-pot procedure described in Figure 3 (**B** and **C**). Briefly, ST3-CMP-Neu5Ac synthetase fusion (42) catalysed the formation of CMP-Neu5Ac quantitatively from 1 equivalent of Neu5Ac and 1 equivalent of CTP. After removal of the fusion protein by membrane filtration (MWCO 10k) a selected galactoside and a recombinant sialyltransferase (Table II) was introduced to produce the desired Neu5Ac-sialoside (19). Multigram-scale synthesis was performed with a typical yield of 70-90% recovery of sialylated products. To synthesize Neu5Gc and KDN derivatives the one-pot system would include another enzymatic reaction in addition to route **B** and **C**. In this respect, mannose

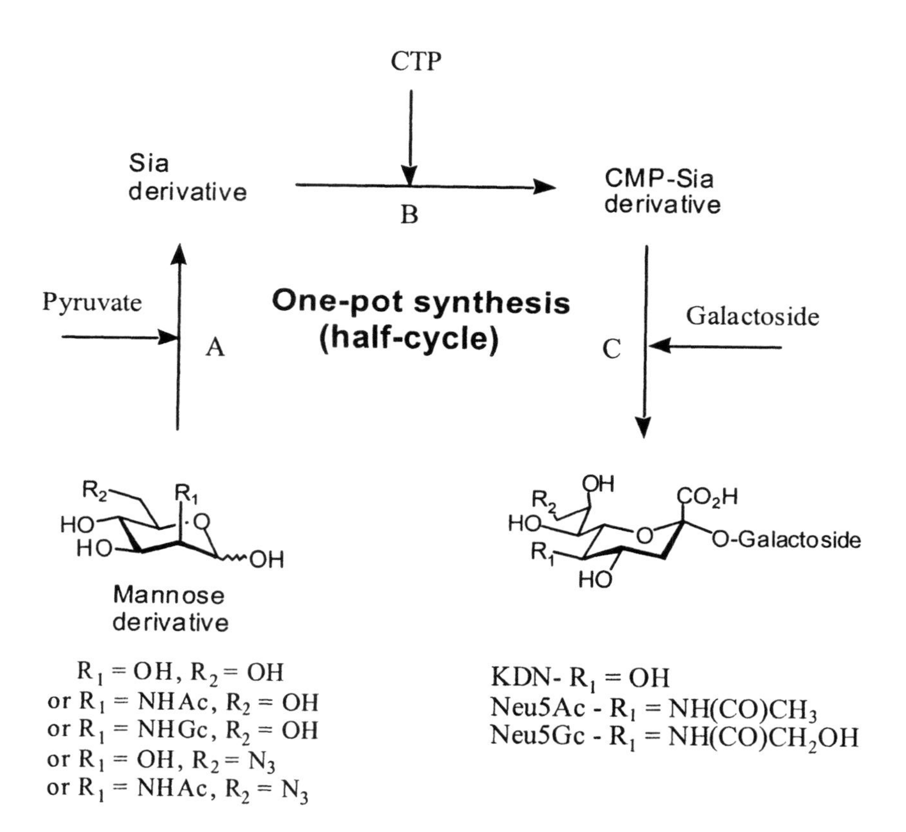

Figure 3. Enzymatic one-pot synthetic procedure of sialyloligosaccharides. ***A*** *Neu5Ac-aldolase,* ***B*** *CMP-Neu5Ac synthetase,* ***C*** *sialyltransferase (adapted from reference 44).*

Table II. Recombinant Sialyltransferases Produced for Synthesis

Sialyltransferase	*Source of Production*	*Produced Activity*[*]
hST6Gal-I	Baculovirus (19)	20
pST3Gal-I	Baculovirus (45)	20
rST3Gal-III	*A. Niger*[#]	50
chST6Gal-I	Baculovirus (46)	10
ST3Gal-Fusion	*E. coli* (42)	6000
ST8 (Cst-II)	*E. coli* (70)	140

NOTE: [#] Unpublished expression construct; [*] Units /L cell culture

derivatives, pyruvate (3 eqv.) and commercial microorganism Neu5Ac aldolase (Toyobo) were introduced into the one-pot half-cycle (Fig. 3, **A**). The enzymes in Table II were able to generate various *N*- and *O*-linked oligosaccharides with α(2-3)-, α(2-6)- or α(2-8)-linked sialic acid derivatives of Neu5Gc, KDN and some of the 9-azido-9deoxy-Neu5Ac-analogs in acceptable yields (45-90%) (44).

O-linked sialyl-oligosaccharides are another class of desired compounds for the biomedical community. These structures are frequently found in various cancer tissues and lymphoma and are highly expressed in many types of human malignancies including colon, breast, pancreas, ovary, stomach, and lung adenocarcinomas (47, 48).

We have previously reported the cloning, expression, and characterization of chicken ST6GalNAc-I and its use in preparative synthesis of the *O*-linked sialoside antigens, STn-, α(2-6)SiaT-, α(2-3)SiaT- and Di-SiaT-antigen (46). Briefly, the recombinant enzyme was expressed in insect cells and purified by CDP-sepharose affinity chromatography to generate approximately 10 U/L of cell culture. The enzymatic activity was evaluated on a set of small acceptor molecules (Table III), and it was found that an absolute requirement for enzymatic activity is that the anomeric position on GalNAc is α-linked to threonine. Thus, *O*-linked sialosides terminating with a protected threonine could successfully be synthesized on a gram-scale reactions (Figure 4). To be able to attach these compounds to other functional groups, the *N*-acetyl protecting group on threonine could be substituted with a biotin derivative before enzymatic extension with chST6GalNAc-I (49). As seen in Table III, the enzyme is not sensitive to bulky groups at this position (compound **6**).

Table III. chST6GalNAc-I Activity of α-D-Galacto Derivatives

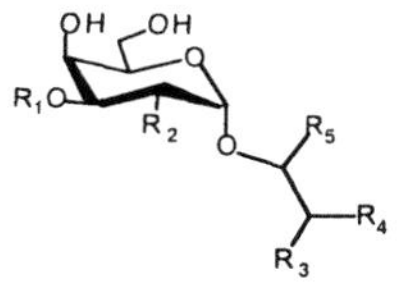

Compound	R_1	R_2	R_3	R_4	R_5	*cpm*	*nmol/ mg x min^{-1}*
D-GalNAc	H	NHAc				0	0.00
1	H	NHAc	N_3	H	H	65	0.06
2	H	NHAc	NHAc	H	H	121	0.11
3c	H	NHAc	NHAc	$COOCH_3$	CH_3	9133	8.60
4	H	N_3	NHAc	$COOCH_3$	CH_3	3043	2.90
5	H	NH_2	NHAc	$COOCH_3$	CH_3	1421	1.30
6	H	NHAc	NHF_{moc}	$COOCH_3$	CH_3	13277	12.50*
7c	Galβ1,3	NHAc	NHAc	$COOCH_3$	CH_3	12760	12.00

NOTE: *Product was isolated by using Sep-Pak (C18) cartridges as described (50).
SOURCE: Adapted from reference (46)

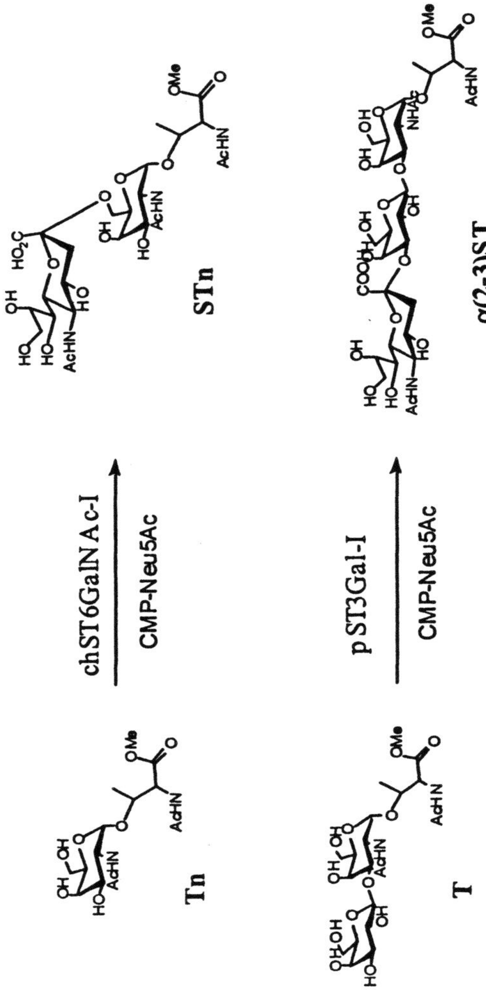

Tn $\xrightarrow[\text{CMP-Neu5Ac}]{\text{chST6GalNAc-I}}$ STn

T $\xrightarrow[\text{CMP-Neu5Ac}]{\text{pST3Gal-I}}$ α(2-3)ST

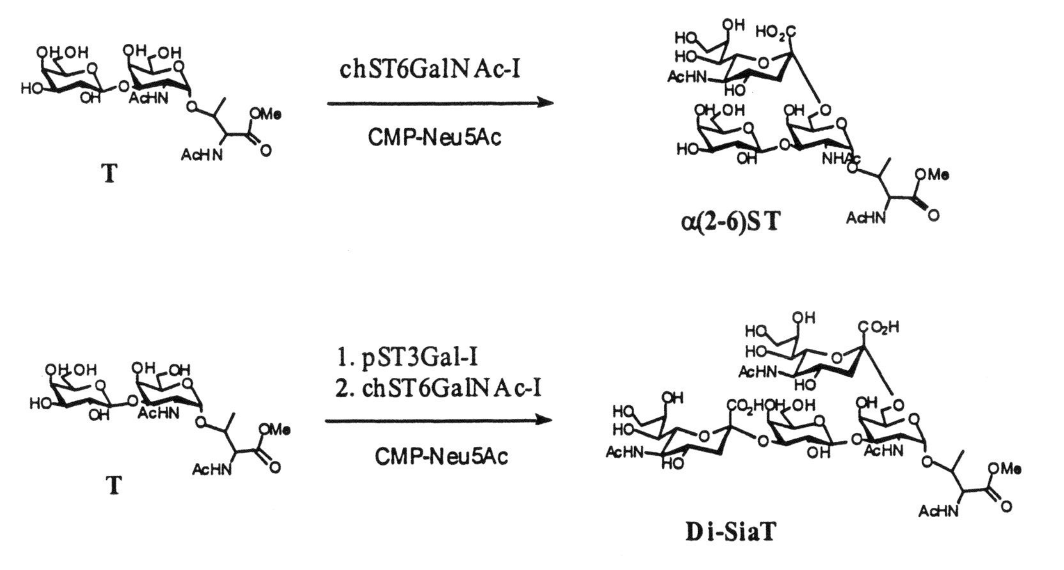

Figure 4. Enzymatic Preparation of O-linked sialosides. Adapted from reference (46)

Synthesis of Ganglioside Mimics

Gangliosides are glycolipids that comprise a structurally diverse set of sialylated molecules. They are attached and enriched in nervous tissues and they have been found to act as receptors for growth factors, toxins and viruses and to facilitate the attachment of human melanoma and neuroblastoma cells (51-57).

Despite the importance of these sialylated ganglioside structures, methods for their efficient preparation have been limiting. The introduction of sialic acid to a glycolipid core structure have shown to be a daunting task, needed complicated engineering with well executed synthetic strategies. Several elegant chemical and a few enzymatic approaches have been reported over decades by various groups to synthesize gangliosides (58-60), and ganglioside analogs (61-69).

Recently, several glycosyltransferase genes from *Campylobacter jejuni* (OH4384) have been identified to be involved in producing various ganglioside-related lipoligosaccharides (LOS) expressed by this pathogenic bacteria (70). Among these genes, cst-II, coding for a bifunctional α(2-3/8) sialyltransferase, have been demonstrated to catalyze transfers of Neu5Ac α(2-3) and α(2-8) to lactose and sialyllactose, respectively. Another gene, cgtA, is coding for a β(1-4)-*N*-acetylgalactosaminyltransferase (β4GalNAcT) that is reported to transfer GalNAc β(1-4) to Neu5Acα(2-3)lactose acceptors generating the GM2 (Neu5Acα(2-3)[GalNAcβ(1-4)]Galβ(1-4)Glc-) epitope. These two glycosyltransferase genes were successfully over expressed in large scale (100 L *E. coli* fermentation) and used in the preparative synthesis of various ganglioside mimics. For synthetic purposes we also conducted an extensive specificity study of these enzymes using neutral and sialylated structures to further specify the synthetic utility of these enzymes (to be published elsewhere).

For a cost-efficient synthesis of GalNAc-containing oligosaccharides, expensive uridine-5'-diphosphate-*N*-acetylgalactosamine (UDP-GalNAc) was produced *in situ* from inexpensive UDP-GlcNAc by the UDP-GlcNAc-4'-epimerase (GalNAc-E). GalNAc-E was cloned from rat liver into the *E. coli* expression vector (pCWori) and expressed in *E. coli* AD202 cells (to be published elsewhere). Briefly, a lactose derivative was elongated with sialic acid repeats using α(2-8)-sialyltransferase and crude CMP-Neu5Ac. Several products (GM3, GD3, GT3) were isolated from this mixture. Increasing CDP-Neu5Ac from 2.5 to 4 equivalents favors the formation of GT3, and minor amounts of GD3 were isolated. Typical yields range from 40-50% of the major compound and 15-20% for the minor compound. Isolated compounds were further furbished with the action of GM2-synthetase (CgtA) and GalE to give the corresponding GM2, GD2, and GT2 structures in quantitative yields (figure 4). Details of the synthesis will be published elsewhere.

Conclusion

The contribution of glycans in different biological processes, e.g. neural development, immunological response, lymphocyte homing and human evolution have been well documented. Yet, the importance of this involvement is under appreciated because of the lack of effective molecular tools to correlate structure with function. In this chapter, we have introduced facile methodologies to build up diverse series of glycans, such as poly-*N*-acetyllactosamine and its corresponding fucosylated and/or sialylated compounds, various sialoside derivatives of *N*- and *O*-linked glycans, and ganglioside mimic structures. Furthermore, a simple route to produce the scarce sialic acid derivatives has been presented. This work demonstrates that chemoenzymatic synthesis of complicated carbohydrate structures can reach a facile and practical level by employing a functional toolbox of different glycosyltransferases. Detailed information of the specificity of these enzymes, indeed, is crucial in developing a library of compounds with an extensive structural assortment. Such a library of carbohydrates is currently being established in our laboratory. This library is being utilized in different frontlines of research in glycobiology. The built-in neutral spacer of the compounds makes them particularly suitable for being immobilized on different matrices. The versatility together with diversity make this library favorable of being used in the newly developed sugar-display technologies in high throughput studies of carbohydrate-protein, as well as, carbohydrate-carbohydrate interactions (71-73). These methodologies can further be employed in constructing non-natural analogs with therapeutic values and entitled to drug discovery.

Acknowledgements

This work was funded by NIGMS and Consortium for Functional Glycomics GM62116. We are thankful to Drs. J. Paulson, W. Wakarchuk and NEOSE Technologies for contributing with enzyme constructs that have been utilized in these studies. We also acknowledge Tokyo Research Laboratories, Kyowa Hakko Kogyo Co. Ltd. for providing bulk quantities of GDP-fucose and UDP-GlcNAc. Special thanks to Dr. D. De Palo for editorial help in preparing this manuscript.

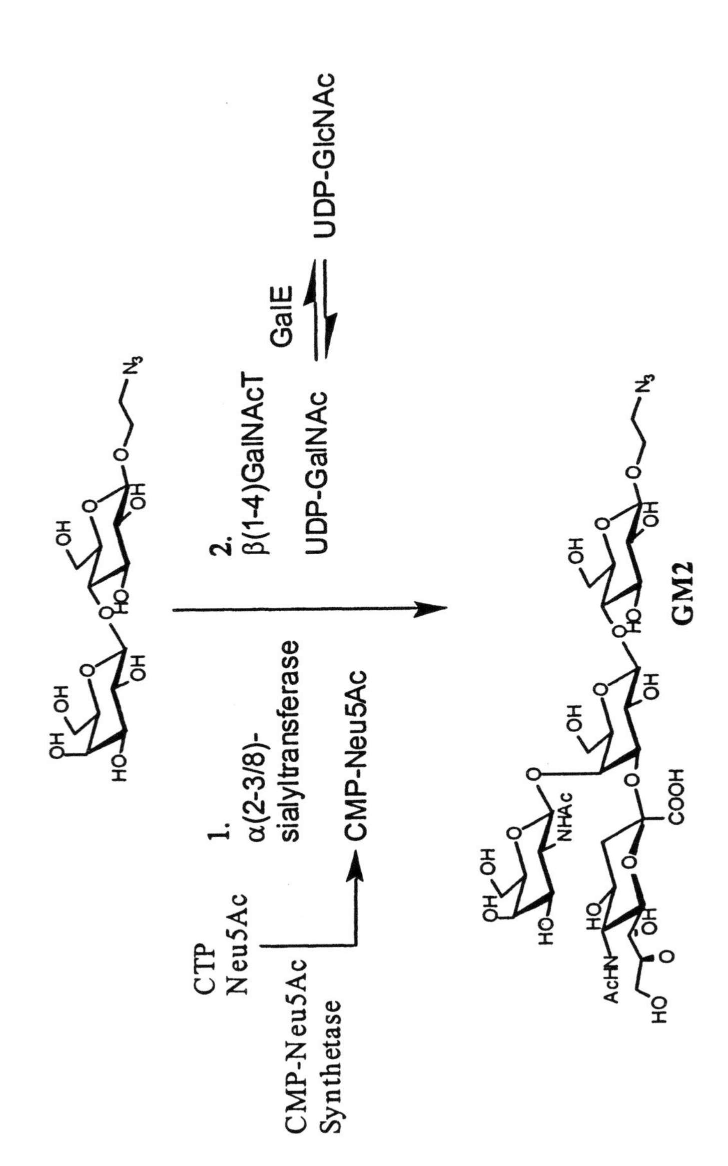
CTP
Neu5Ac
CMP-Neu5Ac
Synthetase
1.
α(2-3/8)-
sialyltransferase
CMP-Neu5Ac
2.
β(1-4)GalNAcT
UDP-GalNAc
GalE
UDP-GlcNAc
NHAc
COOH
AcHN
N_3
GM2

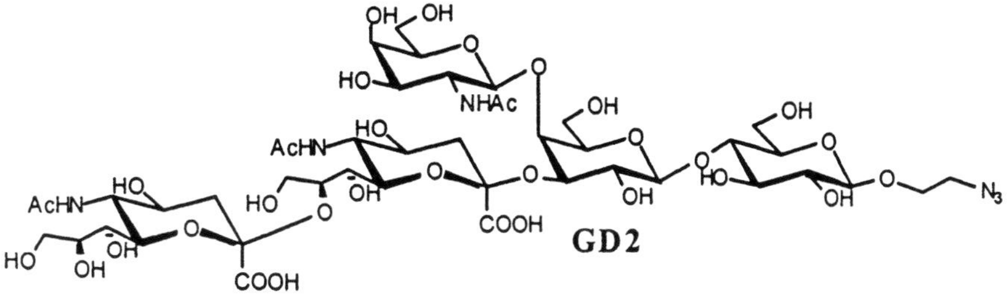

GD2

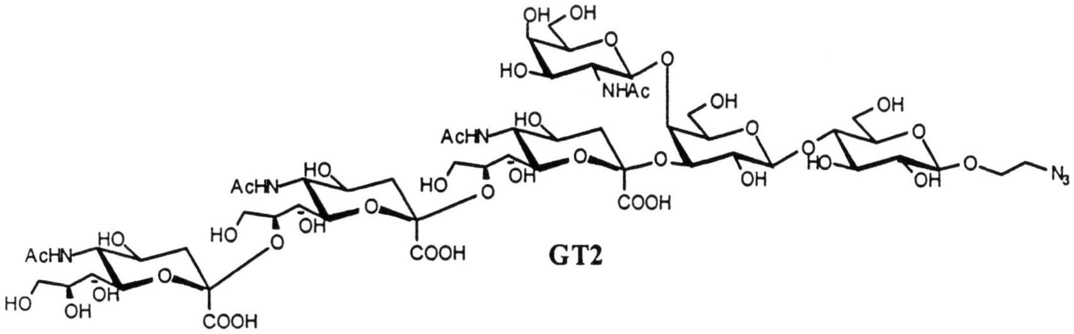

GT2

Figure 5. Synthesis of ganglioside mimics

References

1. Sharon, N. and Lis, H., *Sci. Am.* **1993**, *268*, 82-89.
2. Seeberger, P. H., *J. Carb.Chem.* **2002**, *21*, 613-643.
3. Koeller, K. M. and Wong, C.-H., *Glycobiol.* **2000**, *10*, 1157-1169.
4. Khan, S. H. and Hindsgaul, O., In *Molecular Glycobiology*; Fukuda M. and Hindsgaul O., IRL Press, Oxford, UK, **1994**, 206-229.
5. Toshima, K. and Tatsuta, K., *Chem. Rev.* **1993**, *93*, 1503-1531.
6. Koeller, K. M. and Wong, C.-H., *Nature* **2001**, *409*, 232-240.
7. Wymer, N. and Toone, E. J., *Curr. Opin. Chem. Biol.* **2000**, *4*, 110-119.
8. Gijsen, H. J. M.;Qiao, L.;Fitz, W. and Wong, C.-H., *Chem. Rev.* **1996**, *96*, 443-473.
9. Field, M. C. and Wainwright, L. J., *Glycobiol.* **1995**, *5*, 463-472.
10. Tsuji, S., *J. Biochem.* **1996**, *120*, 1-13.
11. DeAngelis, P. L., *Glycobiol.* **2002**, *12*, 9R-16R.
12. Endo, T. and Koizumi, S., *Curr. Opin. Struct. Biol.* **2000**, *10*, 536-541.
13. Johnson, K. F., *Glycoconj. J.* **1999**, *16*, 141-146.
14. Aly, M. R. E.;Ibrahim, E.-S. I.;El-Ashry, E.-S. H. E. and Schmidt, R. R., *Carbohydr. Res.* **1999**, *316*, 121-132.
15. Ding, Y.;Fukuda, M. and Hindsgaul, O., *Bioorg. Med. Chem. Lett.* **1998**, *8*, 1903-1908.
16. Kretzschmar, G. and Stahl, W., *Tetrahedr.* **1998**, *54*, 6341-6358.
17. Fukuda, M., *Biochim. Biophys. Acta.* **1984**, *780:2*, 119-150.
18. Van den Eijnden, D. H.;Koenderman, A. H. L. and Schiphorst, W. E. C. M., *J. Biol. Chem.* **1988**, *263*, 12461-12471.
19. Blixt, O.;Brown, J.;Schur, M.;Wakarchuk, W. and Paulson, J. C., *J. Org. Chem.* **2001**, *66*, 2442-2448.
20. Blixt, O.;van Die, I.;Norberg, T. and van den Eijnden, D. H., *Glycobiol.* **1999**, *9*, 1061-1071.
21. Ujita, M.;McAuliffe, J.;Hindsgaul, O.;Sasaki, K.;Fukuda, M. N. and Fukuda, M., *J. Biol. Chem.* **1999**, *274*, 16717-16726.
22. Appelmelk, B. J.;Shiberu, B.;Trinks, C.;Tapsi, N.;Zheng, P. Y.;Verboom, T.;Maaskant, J.;Hokke, C. H.;Schiphorst, W. E. C. M.;Blanchard, D.;SimoonsSmit, I. M.;vandenEijnden, D. H. and Vandenbroucke Grauls, C. M. J. E., *Infect. Immun.* **1998**, *66*, 70-76.
23. Leppaenen, A.;Penttilae, L.;Renkonen, O.;McEver, R. P. and Cummings, R. D., *J. Biol. Chem.* **2002**, *277*, 39749-39759.
24. Renkonen, O., *Cell. Mol Life Sci.* **2000**, *57*, 1423-1439.
25. Baldus, S. E.;Zirbes, T. K.;Weingarten, M.;Fromm, S.;Glossmann, J.;Hanisch, F. G.;Monig, S. P.;Schroder, W.;Flucke, U.;Thiele, J.;Holscher, A. H. and Dienes, H. P., *Tumor Biology.* **2000**, *21*, 258-266.
26. Cho, M. and Cummings, R. D., *TIGG..* **1997**, *9*, 47-56, 171-178.

27. Murray, B. W.;Takayama, S.;Schultz, J. and Wong, C. H., *Biochem.* **1996**, *35*, 11183-11195.
28. Weston, B. W.;Nair, R. P.;Larsen, R. D. and Lowe, J. B., *J. Biol. Chem.* **1992**, *267*, 4152-4160.
29. Kimura, H.;Shinya, N.;Nishihara, S.;Kaneko, M.;Irimura, T. and Narimatsu, H., *Biochem. Biophys. Res. Comm.* **1997**, *237*, 131-137.
30. Chandrasekaran, E. V.;Jain, R. K.;Larsen, R. D.;Wlasichuk, K. and Matta, K. L., *Biochem.* **1996**, *35*, 8914-8924.
31. Devries, T.;Vandeneijnden, D. H.;Schultz, J. and Oneill, R., *FEBS Lett.* **1993**, *330*, 243-248.
32. Devries, T. and van den Eijnden, D. H., *Biochem.* **1994**, *33*, 9937-9944.
33. Angata, T. and Varki, A., *Chem Rev.* **2002**, *102*, 439-470.
34. Schauer, R., *Trends in Biochem. Sci.* **1985**, *10*, 357-360.
35. Varki, A., *Glycobiol.* **1993**, *3*, 97-130.
36. Lasky, L. A., *Ann. Rev. Biochem.* **1995**, *64*, 113-139.
37. Varki, A., *Proc. Natl. Acc. Sci.* **1994**, *91*, 7390-7397.
38. Crocker, P. R., *Curr. Opin. Struct. Biol.* **2002**, *12*, 609-615.
39. Halcomb, R. L. and Chappell, M. D., *J. Carb. Chem.* **2002**, *21*, 723-768.
40. Boons, G.-J. and Demchenko, A. V., *Chem. Rev.* **2000**, *100*, 4539-4565.
41. Hideharu, I. and Makoto, K., *TIGG.* **2001**, *13*, 57-64.
42. Gilbert, M.;Bayer, R.;Cunningham, A.-M.;DeFrees, S.;Gao, Y.;Watson, D. C.;Young, N. M. and Wakarchuk, W. W., *Nature Biotechnol.* **1998**, *16*, 769-772.
43. Ichikawa, Y.;Look, G. C. and Wong, C. H., *Anal. Biochem.* **1992**, *202*, 215-238.
44. Blixt, O. and Paulson, J. C., *Adv. Synth. Catal.* **2003**, *345*, 687-690.
45. Williams, M. A.;Kitagawa, H.;Datta, A. K.;Paulson, J. C. and Jamieson, J. C., *Glycoconj. J.* **1995**, *12*, 755-761.
46. Blixt, O.;Allin, K.;Pereira, L.;Datta, A. and Paulson, J. C., *J. Am. Chem. Soc.* **2002**, *124*, 5739-5746.
47. Dabelsteen, E., *J. Pathol.* **1996**, *179*, 358-369.
48. Itzkowitz, S. H.;Yuan, M.;Montgomery, C. K.;Kjeldsen, T.;Takahashi, H. K. and Bigbee, W. L., *Cancer Res.* **1989**, *49*, 197-204.
49. Blixt, O.;Collins, B. E.;Van Den Nieuwenhof, I. M.;Crocker, P. R. and Paulson, J. C., *Manuscript in press* (**2003** *J. Biol. Chem.* Manuscript M304331200).
50. Palcic, M. M.;Heerze, L. D.;Pierce, M. and Hindsgaul, O., *Glycoconj. J.* **1988**, *5*, 49-63.
51. Kiso, M., *Nippon Nogei Kagaku Kaishi.* **2002**, *76*, 1158-1167.
52. Gagnon, M. and Saragovi, H. U., *Expert Opinion on Therapeutic Patents.* **2002**, *12*, 1215-1223.
53. Svennerholm, L., *Adv. Gen.* **2001**, *44*, 33-41.

54. Schnaar, R. L., *Carbohydr. Chem. Biol.* **2000**, *4*, 1013-1027.
55. Ravindranath, M. H.;Gonzales, A. M.;Nishimoto, K.;Tam, W.-Y.;Soh, D. and Morton, D. L., *Ind. J. Exp. Biol.* **2000**, *38*, 301-312.
56. Rampersaud, A. A.;Oblinger, J. L.;Ponnappan, R. K.;Burry, R. W. and Yates, A. J., *Biochem. Soc. Trans..* **1999**, *27*, 415-422.
57. Nohara, K., *Seikagaku.* **1999**, *71*, 337-341.
58. Ito, Y., *Kagaku to Seibutsu.* **1991**, *29*, 788-797.
59. Kiso, M. and Hasegawa, A., *Yukagaku.* **1991**, *40*, 370-378.
60. Hasegawa, A. and Kiso, M., *Carbohydrates.* **1992**, 243-266.
61. Hossain, N.;Zapata, A.;Wilstermann, M.;Nilsson, U. J. and Magnusson, G., *Carbohydr. Res.* **2002**, *337*, 569-580.
62. Bernardi, A.;Boschin, G.;Checchia, A.;Lattanzio, M.;Manzoni, L.;Potenza, D. and Scolastico, C., *Eur. J. Org. Chem..* **1999**, 1311-1317.
63. Ito, H.;Ishida, H.;Kiso, M. and Hasegawa, A., *Carbohydr. Res.* **1998**, *306*, 581-585.
64. Ando, H.;Ishida, H.;Kiso, M. and Hasegawa, A., *Carbohydr. Res.* **1997**, *300*, 207-217.
65. Castro-Palomino, J. C.;Ritter, G.;Fortunato, S. R.;Reinhardt, S.;Old, L. J. and Schmidt, R. R., *Angew. Chem. Int. Ed.* **1997**, *36*, 1998-2001.
66. Castro-Palomino, J. C.;Simon, B.;Speer, O.;Leist, M. and Schmidt, R. R., *Eur. J. Org. Chem.* **2001**, *7*, 2178-2184.
67. Ito, Y. and Paulson, J. C., *J. Am. Chem. Soc.* **1993**, *115*, 1603-1605.
68. Duclos, R. I., *Carbohydr. Res.* **2000**, *328*, 489-507.
69. Liu, K. K. C. and Danishefsky, S. J., *J. Am. Chem. Soc.* **1993**, *115*, 4933-4934.
70. Gilbert, M.;Brisson, J.-R.;Karwaski, M.-F.;Michniewicz, J.;Cunningham, A.-M.;Wu, Y.;Young, N. M. and Wakarchuk, W. W., *J. Biol. Chem.* **2000**, *275*, 3896-3906.
71. Fazio, F.;Bryan, M. C.;Blixt, O.;Paulson, J. C. and Wong, C.-H., *J. Am. Chem. Soc.* **2002**, *124*, 14397-14402.
72. Bidlingmaier, S. and Snyder, M., *Chem. Biol.* **2002**, *9*, 400-401.
73. Houseman, B. T. and Mrksich, M., *Chem. Biol.* **2002**, *9*, 443-454.

Chapter 8

Artificial Golgi Apparatus: Direct Monitoring of Glycosylation Reactions on Automated Glycosynthesizer

Shin-Ichiro Nishimura[1,2,*], Noriko Nagahori[1], Reiko Sadamoto[1], Kenji Monde[1], and Kenichi Niikura[1]

[1]Laboratory for Bio-Macromolecular Chemistry, Division of Biological Sciences, Graduate School of Science, Hokkaido University, N21 W11, Kita-ku, Sapporo 001–0011, Japan
[2]Glycochemosynthesis Team, Research Center for Glycoscience, National Institute of Advanced Industrial Science and Technology (AIST), Sapporo 062–8517, Japan
[*]Corresponding author: Telephone: +81–11–706–9043, Fax: +81–11–706–90–9042, email: shin@glyco.sci.hokudai.ac.jp

Enzymatic synthesis is potential method for the construction of glycoconjugates as well as conventional chemical synthesis. Use of glycosyltransferases in the synthetic schemes has accelerated practical synthesis of oligosaccharides and related compounds. However, there is no systematic approach for establishing "automated glycosynthesizer" based on the combined chemical and enzymatic strategy. We have found that water-soluble glycopolymers bearing multivalent sugars become excellent glycosyl acceptor substrates of glycosyltransferases. Affinity of enzymes with multivalent glycosyl acceptors was drastically enhanced by polymeric sugar-cluster effect. Some designed linkers that can be recognized by specific enzyme permitted release of products from polymer supports under mild and selective conditions.

By employing these water-soluble primers with some engineered glycosyltransferases, we succeeded to produce "artificial Golgi apparatus", an automatic glycosynthesizer controlled by computer. In this chapter, our recent approach for direct monitoring of sugar -elongation reactions by using surface plasmon resonance (SPR) method will be described, since direct and real time monitoring of enzymatic carbohydrate synthesis is strongly required for materializing practically usable artificial Golgi apparatus.

Real-time monitoring of enzymatic glycosylation processes is desirable not only for establishing efficient automated synthetic methods of various carbohydrates (1) but also for understanding the catalytic mechanism of glycosyltransferases (2,3). In addition, high throughput assay is also required for screening novel inhibitors or modulators of glycosyltransferases among a numerous number of combinatorial libraries in pharmaceutical fields. Although a few practical methods for continual monitoring of glycosyltransferase activity in aqueous solutions such as spectrophotometric assay using pyruvate kinase-based reaction (4) and fluorescence resonance energy transfer between donor and acceptor substrates (5) were reported, these methods are not suited for direct monitoring of the sugar elongation reactions by glycosyltransferases immobilized on the polymer supports (6). From the viewpoint of automated synthesis on some solid phase materials, such spectrophotometric methods as well as NMR and MALDI-TOF mass spectroscopy (7) can not be used as facile and efficient tools for direct and real-time monitoring of the progress of glycosylation reactions.

In the course of our research on the enzyme-assisted synthesis of glycoconjugates (6,8), we have employed some glycosyltransferases expressed as fusion proteins with maltose binding protein (MBP) in order to immobilize these enzymes on the surface of polymer supports having maltotriose branches. Since MBP-combined enzymes exhibited highly specific and strong affinity with maltooligosaccharide carrying materials, it was suggested that glycosyltransferases can be simply immobilized on the basis of "specific carbohydrate-protein interaction" without any reagent to crosslink enzymes with supporting materials (9) (Figure 1). These findings prompted us to apply this efficient immobilization process for preparing highly-oriented glycosyltransferase microarray on a biosensor chip used in a SPR technique. Here we describe a novel method for the real-time monitoring of sugar elongation reactions catalyzed by galactosyltransferase microarray displayed on the surface of glycolipid LB thin films (10).

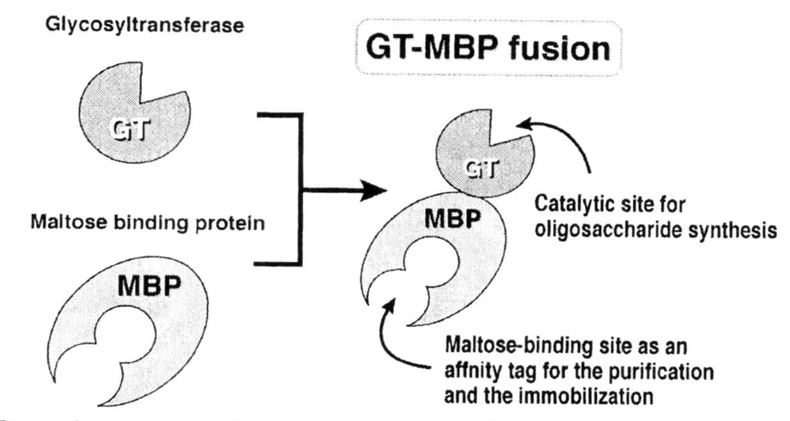

Figure 1. A concept of the engineered glycosyltransferases by means of fusion protein with maltose-binding protein

Results and Discussion

Although some self-assembled monolayers of alkanethiolates on gold have been employed as a platform for immobilizing carbohydrates (11), we have designed and synthesized a novel type of photopolymerizable glycolipids that will be widely applied for the preparation of carbohydrate-based microarrays (12). Scheme 1 indicated a basic concept for the construction of a glycosyltransferase microarray displayed on the LB membrane. Firstly, the monolayer prepared from a photopolymerizable glyceroglycolipid **1** with a matrix lipid **2** (Figure 2) was polymerized by UV irradiation at 254 nm to produce the polydiacetylene thin film (step i and ii) according to the previous reports (13). Then, the glycopolydiacetylene membrane was transferred onto a glass plate or a SPR sensor chip (step iii). Finally, these glass plate and sensor chip were immersed into MBP-GalT solution to immobilize a fusion protein through the specific interaction between MBP and maltotriose residue.

Advantage of the use of the polydiacetylene-type glycolipid LB membrane is that polymerized films prevent a collapse of the monolayer by flip-flop of natural glycolipids (14) when they are exposed into the air during the transferring LB membrane to plate or sensor chip. In addition, polymerization would provide a stable platform for MBP-GalT (a guest fusion protein) on the sensor chip and avoid the dissociation of the glycolipid-MBP-GalT complex from the film. Here, diacetylene-containing maltotriose-carrying glycolipid **1** was designed and synthesized for the immobilization of MBP-GalT via specific maltotriose-MBP interaction (Scheme 2). Synthetic details are described in the experimental section. Diacetylene-containing phosphatidylcholine **2** was selected as the matrix lipid because phosphatidyl choline is an abundant lipid in biomembrane and often suppresses the nonspecific adsorption of proteins (15).

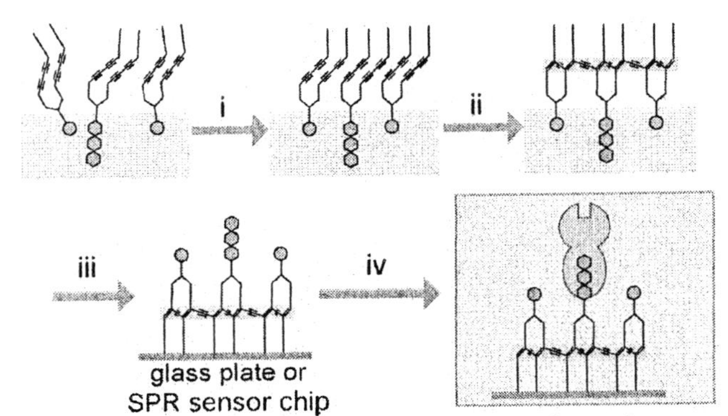

Scheme 1. Procedure to construct a glycosyltransferase biosensorchip. (i) formation of a monolayer, (ii) photopolymerization by UV irradiation, (iii) transfer of the polymerized membrane onto a solid surface, (iv) immobilization of MBP-GalT through maltotriose-MBP interaction.

Figure 2. Chemical structures of photopolymerizable lipids for the preparation of LB films and acceptor substrates for the sugar elongation reaction by MBP-GalT.

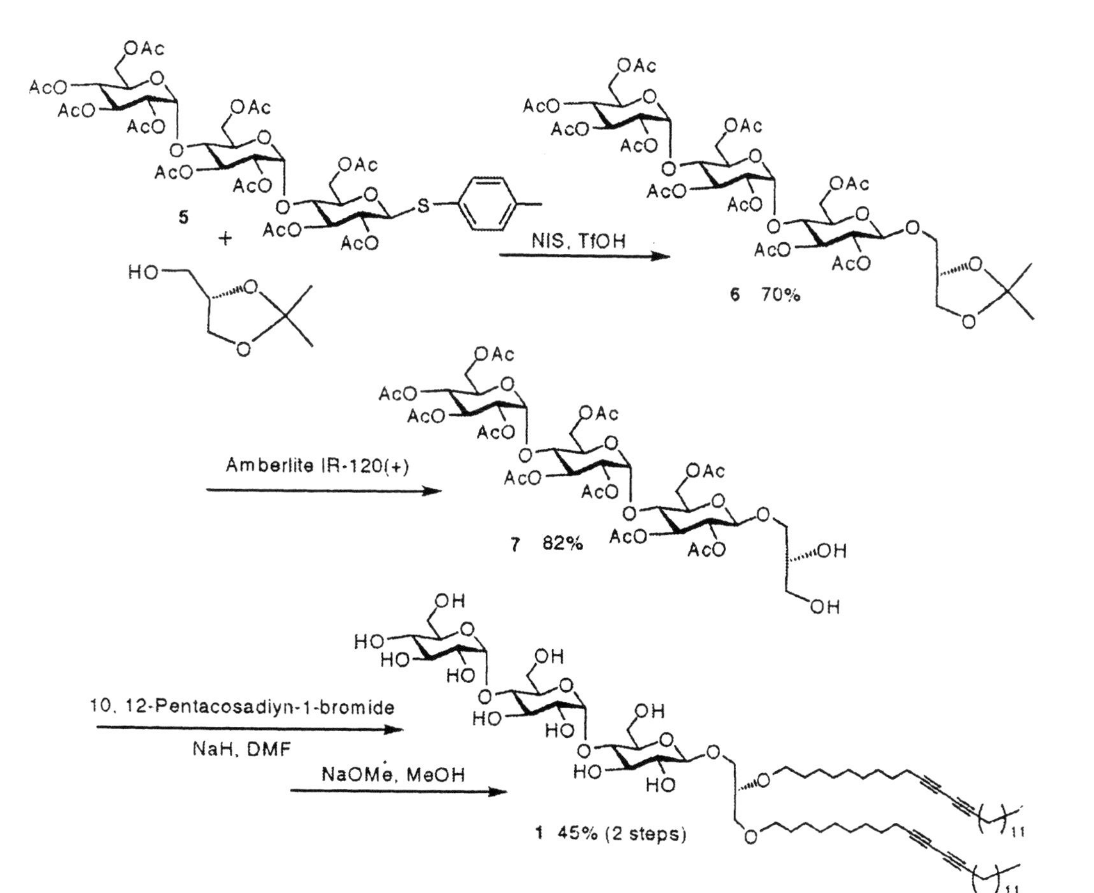

Scheme 2. Synthesis of photopolymerizable glycolipid 1

The morphology of MBP-GalT bound on the surface of LB films was observed by using atomic force microscope (AFM). Figure 3 shows the AFM images of MBP-GalT adsorbed on the glycopolydiacetylene films prepared from monolayers with various lipid compositions of **1** and **2**. The AFM images exhibited the formation of the specific protein arrays such as networks and dendrites, depending on the carbohydrate density in the LB film. It was also suggested from the section views that MBP-GalT formed a single layer (monolayer) on the glycolipid LB membrane in all cases. When MBP-GalT was incubated with the film composed of the matrix lipid **2** alone as a control experiment (photo a), only a small amount of protein was adsorbed on the film, indicating that MBP-GalT was bound to the glycolipid LB membrane by the specific carbohydrate-protein interaction between maltotriose residue and MBP-GalT. LB films containing 5% (photo b), 10% (photo c), or 20% (photo d) of **1** were covered by much larger amount of protein molecules than the film prepared with 100 % glycolipid **1** (photo e) [data shown partly in Figure 4(b)]. It seems that there is an appropriate sugar density to immobilize protein efficiently.

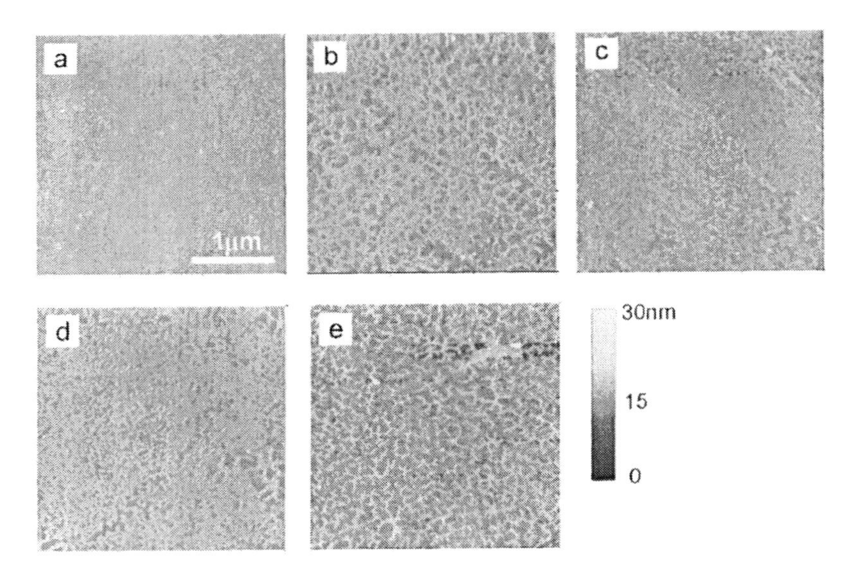

*Figure 3. AFM images of MBP-GalT immobilized on the glycolipid LB films. The density of glycolipid **1** in the matrix lipid **2** was (a) 0 %, (b) 5 %, (c) 10 %, (d), 20 % and (e) 100%.*

Immobilized MBP-GalT on gold sensor chips (ratio of glyceroglycolipid **1** : matrix lipid **2** = 1 : 4) were employed for further activity evaluation study by SPR (Biacore-X, Biacore AB) with the addition of the solutions containing glycosyl donor (UDP-Gal) and/or acceptor substrates (compound **3** or **4**). Figure 4 (red solid line) shows a typical SPR sensorgram demonstrating galactose transfer reaction from UDP-Gal to acceptor polymer **3** catalyzed by MBP-GalT on a sensor chip. In this experiment, the glycosyl acceptor **3** was added at the point arrowed by (A) and the flow cell was rinsed with buffer solution at the time arrowed by (B). When the polymer **3** was injected in the presence of 0.5 mM UDP-Gal, a large sigmloidal response (2300 RU) was recorded (red solid line). The value of 2300 RU seems to be reasonable when it is assumed that a single layer of the polymer coverers the sensor surface. On the other hand, no response was monitored when the same amount of acceptor **3** was injected in the absence of UDP-Gal (blue solid line).

When a monomeric GlcNAc derivative **4** was injected with UDP-Gal, negligibly low affinity was detected (green dotted line). These results correspond with our previous observation that acceptor sugar residues highly branched on the water soluble polymers become excellent substrates for enzymatic glycosylation reactions due to a sort of "cluster effect" (6,16). Moreover, the large sigmoidal response of SPR found in the presence of both UDP-Gal and polymeric acceptor **3** suggests clearly that activation of GalT by the predominant binding with UDP-Gal may be an essential step for successful galactosylating process to the acceptor substrate as illustrated in Figure 4b. Recently, X-ray crystallographic analysis of the complex of GalT with UDP-Gal also showed a significant conformational change and this crucial step may be necessary for creating a properly arranged acceptor-binding pocket (3).

To investigate the effect of the sugar density of the LB film on the activity of MBP-GalT immobilized on the sensor chip, sensorgrams observed in the films prepared by using some different monomer ratio were compared as shown in Figure 5. In all cases, characteristic sigmoidal curves were obtained. The maximum intensity of sensorgrams and the protein coverage estimated by AFM images were co-plotted as a function of the sugar density on the film. Interestingly, enzyme activity detected by SPR analyses was strongly dependent on the amount of immobilized proteins as indicated by the surface area covered by this engineered enzyme. Our attention is now directed toward the versatility of the present GalT-immobilized sensor chip to an automated synthetic system of glycoconjugates.

NMR spectrum of the product exhibited a perfect glycosylation on the designated polymer primer **3**. This result suggests that the present method to use novel type of glycosyltransferase microarray chip might materialize "automated microsynthesizer" for glycoconjugates.

Conclusion

β(1→4) Galactosyltransferase expressed as a fusion protein with maltose binding protein (MBP-GalT) was displayed specifically on Langmuir-Blodgett (LB) membrane prepared by photopolymerization of maltotriose-carrying glycolipid with 1,2-bis(10,12-tricosadiynoyl)-*sn*-glycero-3-phosphocholine.

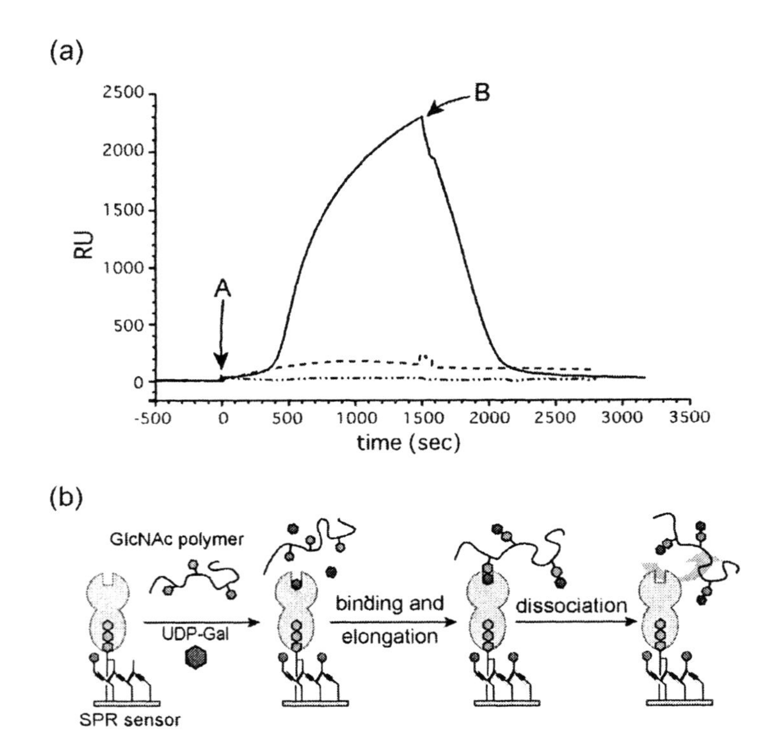

*Figure 4. (a) SPR sensorgram upon adding of glycosyl donor and/or acceptor substrates. A, injection of the substrate mixture. B, washing the flow cell with buffer solution. The mixture of acceptor **3** (42 μM based on GlcNAc) and UDP-Gal (0.5 mM) was injected (solid line), or acceptor **3** alone (– – –), or acceptor **4** (42 μM) with UDP-Gal (– – · · – –). (b) The plausible process of galactosylation on the SPR sensor chip, suggested from the sensorgram.*

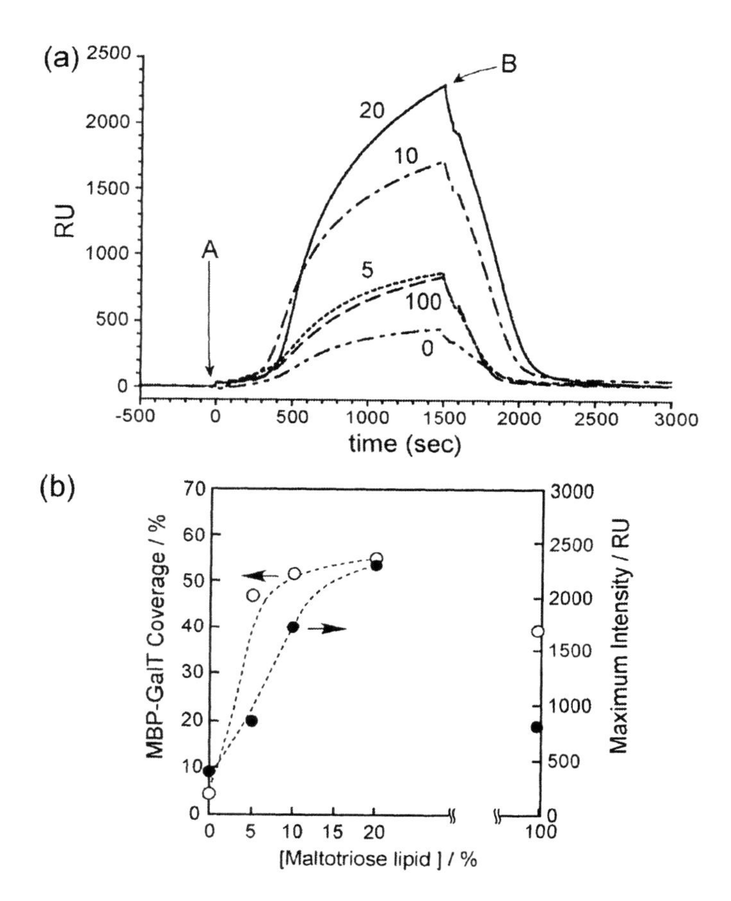

Figure 5. The effect of the sugar density of the LB film on the activity of immobilized MBP-GalT.

*(a) SPR sensorgram upon addition of the mixture of acceptor polymer **3** (42 μM based on GlcNAc) and donor UDP-Gal (0.5 mM). The density of glycolipid **1** in LB film was 0, 5, 10, 20 or 100 % as shown in the figure. (b) Correlation between the area occupied with MBP-GalT and the maximum intensity obtained by SPR measurement. The area was calculated on the computer simulation. First all AFM images were rearranged to a black and white image with 800 x 800 dots, then dots were substituted by "0" for black and by "1" for white. The protein area corresponds to the total number of "1".*

Catalytic activity of MBP-GalT on the LB film was directly monitored by surface plasmon resonance (SPR) method using GlcNAc-carrying water-soluble polymer as an acceptor substrate. Highly sensitive sigmoidal-type signals were obtained upon an addition of the acceptor substrate in the presence of donor substrate, UDP-galactose (UDP-Gal), while the binding of acceptor substrate was not detected in the absence of UDP-Gal. The intensities of the signals were dependent on the amount of immobilized MBP-GalT on the LB film, which was estimated from the images obtained by atomic force microscope (AFM). The simple and efficient monitoring method described herein will be applied for other glycosyltransferase-MBP fusion proteins and this approach may accelerate the establishment of the practical automated glycosynthesis on the basis of enzymatic reactions.

Experimental

Synthesis of photopolymerizable glyceroglycolipid (1):

Sodium hydride in oil (60%, 40 mg) was added to a mixture of 10,12-pentacosadiyn-1-bromide (300 mg, 0.71 mmol) and compound **7** (200 mg, 0.2 mmol) in an anhydrous DMF. The mixture was stirred for 2 days at room temperature, and concentrated in *vacuo*. The residue was dissolved in CH_2Cl_2, washed with water and brine. The solvent was evaporated and the residue was dissolved in dry methanol. Sodium methoxide (10.8 mg, 0.2 mmol) was added to the solution and stirred for 12 hr. After neutralized with Amberlite IR-120 (H+) resin, the solvent was evaporated. The residue was purified by silica gel column chromatography (4:1 $CHCl_3$-MeOH) to give compound **1** as white solid in 45% yield. ^{1}H NMR ($CDCl_3/CD_3OD$ (7/3), 500 MHz) δ = 5.12 (d, 1H, J = 3.47 Hz), 5.11 (d, 1H, J = 3.47 Hz), 4.32 (d, 1H, J = 7.83 Hz), 3.95-3.20 (m, 17H), 3.09 (dd, 1H, J =7.78, 9.77 Hz), 2.22 (t, 8H, J = 8.20), 1.65-1.20 (m, 64H), 0.85 (t, 6H, J = 6.93 Hz). ESI-Mass (pos) Calcd for $C_{71}H_{122}O_{18}Na$: m/z 1285.8 $[M+Na]^+$, Found m/z 1285.6 $[M+Na]^+$.

Preparation of GalT microarray

Glycolipid **1** and matrix lipid **2** at various concentrations (**1** / **2** = 1 / 0, 5 / 1, 9 / 1, 19 / 1 and 0 / 1) in chloroform (10^{-3}M ~ 10^{-4}M) was spread on the surface of an aqueous subphase. After waiting for 10 min to vaporize the chloroform, the mixed lipids were compressed to the pressure of 30 mN/m at a rate of 0.1 nm^2/molecule/min to form a monolayer. The monolayer was equilibrated for 10 min under the same surface pressure and polymerized by UV irradiation (254 nm, 8 W) for 10 min. The distance between the lamp and the monolayer surface was 12 cm. The polymerized monolayer was transferred onto the OTS (octadecyltrichlorosilane) coated glass plate or the SPR sensor chip carrying monolayer of alkanethiol (*n*-octadecyl mercaptan) using horizontal deposition technique. Then the glass plate carrying polymerized glycolipid film was placed on the buffer solution containing 0.01 mg/mL MBP-GalT (50 mM Tris-HCl, pH 7.4) for 1 h. The surface of the glass plate was rinsed with buffer solution and pure water (Milli-Q) for AFM observation.

Monitoring the substrate binding to the MBP-GalT array on SPR sensor chip

SPR sensor chips, of which one side is covered with gold, were immersed into an ethanol solution of 1 mM *n*-octadecyl mercaptan for 12 h. The hydrophobic surface was washed with ethanol and benzene and dried under nitrogen-stream. MBP-GalT was bound to the glycolipid LB films of **1** / **2** (1 / 0, 5 / 1, 9 / 1, 19 / 1 and 0 / 1) on the modified SPR sensor chips. The sensor chip adsorbed MBP-GalT was set to the SPR instrument. The sensor chip was equilibrated with 10 mM Tris-HCl buffer containing 150 mM NaCl and 10 mM $MnCl_2$. The buffer solution containing substrates for the enzyme was injected to the MBP-GalT immobilized SPR sensor chip and the signal was recorded. The concentrations of glycosyl donor (UDP-Gal) and glycosyl acceptor (compound **3** or **4**) substrates were 0.5 mM and 0.042 mM respectively.

Selected spectral data for the product:

^{1}H NMR (D_2O, 600 MHz) δ = 7.25 (m, 5H, Ph), 4.7 (D_2O and Phe-αH), 4.43 (m, 2H, H-1 and H-1'), 3.86-3.80 (d, 4H, H-4, H-2, H-4' and H-2'), 3.70 (dd, 1H, H-3'), 3.60 (t, 1H, H-5'), 3.48 (m, 2H, OCH_2), 3.37 (2, 2H, H-6), 3.12-2.94 (br, 10H, H-3', H-5', H-6' and NCH_2 and $PhCH_2$), 2.27-2.13 (m, 19H), 1.96 (s, 3H, $NHCH_3$), 1.70-1.04 (m, 47H).

Acknowledgments

This work was supported by a grant for the Glycocluster Project from NEDO. We also thank Ms. A. Maeda, Ms. H. Matsumoto, and Ms. S. Oka of the Center of Instrumental Analysis, Hokkaido University, for measuring elemental analysis and mass spectroscopy data.

References

(1) (a) P. Sears, C. H. Wong, *Science* 2001, *291*, 2344-2350; (b) S. Nishimura, *Curr. Opin. Chem. Biol.* 2001, *5*, 325-335; (c) O. J. Plante, E. R. Palmacci, P. H. Seeberger, *Science* 2001, *291*, 1523-1527.

(2) Y. Nishikawa, W. Pegg, H. Paulsen, H. Schachter, *J. Biol. Chem.* 1988, *263*, 8270-8281.

(3) B. Ramakrishnan, P. K. Qasba, *J. Mol. Biol.* 2001, *310*, 205-218.

(4) D. K. Fitzgerald, B. Colvin, R. Mawal, K. E. Ebner, *Anal. Biochem.* 1970, *36*, 43-61.

(5) K. Washiya, T. Furuike, F. Nakajima, Y. C. Lee, S. I. Nishimura, *Anal. Biochem.* 2000, *283*, 39-48.
(6) S. Nishiguchi, K. Yamada, Y. Fuji, S. Shibatani, A. Toda, S.-I. Nishimura, *Chem. Commun.* 2001, 1944-1955.
(7) H. Ando, S. Manabe, Y. Nakahara, Y. Ito, *J. Am. Chem. Soc.* 2001, *123*, 3848-3849.
(8) (a) K. Yamada, E. Fujita, S. Nishimura, *Carbohydr. Res.* 1997, *305*, 443-461; (b) K. Yamada, S. Matsumoto, S.-I. Nishimura, *Chem. Commun.* 1999, 507-508; (c) S.-I. Nishimura, K. Yamada, *J. Am. Chem. Soc.* 1997, *119*, 10555.
(9) A. Toda, K. Yamada, S.-I. Nishimura, *Adv. Synth. Catal.* 2002, *344*, 61-69.
(10) N. Nagahori, K. Niikura, R. Sadamoto, M. Taniguchi, A. Yamagishi, K. Monde, S. –I.Nishimura, *Adv. Synth. Catal.* 2003, *345*, 729-734.
(11) (a) M. J. Hernaz, J. M. Fuente, A. G. Barrientos, S. Penades, *Angew. Chemm. Int. Ed.* 2002, *41*, 1554-1557; (b) B. T. Houseman, M. Mrksich, *Chem. Biol.* 2002, *9*, 443-454; (c) M. C. Bryan, O. Plettenburg, P. Sears, D. Rabuka, S. Wacowich-Sgarbi, C. H. Wong, *Chem. Biol.* 2002, *9*, 713-720.
(12) N. Nagahori, K. Niikura, R. Sadamoto, K. Monde, S.-I. Nishimura, *Aust. J. Chem.* 2002, *56*, 567-576.
(13) (a) D. H. Charych, J. O. Nagy, W. Spevak, M. D. Bednarski, *Science* 1993, *261*, 585-588; (b) D. Charych, Q. Cheng, A. Reichert, G. Kuziemko, M. Stroh, J. O. Nagy, W. Spevak, R. C. Stevens, *Chem. Biol.* 1996, *3*, 113-120.
(14) M. W. Charles, *J. Colloid. Interface. Sci.* 1969, *31*, 569-572.
(15) T. Hasegawa, Y. Iwasaki, K. Ishihara, *Biomaterials* 2001, *22*, 243-251.
(16) R. T. Lee, Y. C. Lee, *Glycoconj. J.* 2000, *17*, 543-551.

Chapter 9

Sugar Engineering with Glycosaminoglycan Synthases

Paul L. DeAngelis

Department of Biochemistry and Molecular Biology, Oklahoma Center for Medical Glycobiology, University of Oklahoma Health Sciences Center, 940 S. L. Young Boulevard, Oklahoma City, OK 73104

The chemoenzymatic synthesis of natural and artificial glycosaminoglycans [GAGs] is now possible with the advent of three cloned, recombinant *Pasteurella* GAG synthases. Depending on the choice of the catalyst, the acceptor molecule and the UDP-sugar precursors, a wide variety of potentially useful materials based on the GAGs hyaluronan, chondroitin, and/or N-acetylheparosan may be created. We present the syntheses of (*a*) short monodisperse oligosaccharides, (*b*) hybrid polymers composed of two disparate GAGs, (*c*) GAGs immobilized on surfaces, (*d*) libraries of GAG oligosaccharides, and (*e*) novel GAGs containing unnatural functionalities. These materials should have many future biomedical applications due to the prevalence and the importance of GAGs in the mammalian body.

Chemoenzymatic Synthesis of Sugar Polymers

Naturally occurring oligosaccharides and polysaccharides from various organisms are often difficult to prepare in a pure, defined, monodisperse form. Sugar polymers, especially molecules with chain lengths longer than five monosaccharides, are also difficult to synthesize by strictly organic synthesis. In contrast, chemoenzymatic synthesis offers the potential to harness enzyme catalysts for rapid, efficient reactions. The exquisite stereochemical and regioselective control of glycosyltransferases allows the preparation of target compounds without multistep reactions involving protection/deprotection strategies. The **major obstacles** to utilizing the chemoenzymatic approach are *i*) identifying or constructing suitable enzymes, *ii*) defining and optimizing their synthetic capabilities, and *iii*) harnessing the catalysts at the preparative scale.

Bacterial Capsules and Glycosaminoglycans

Pathogenic bacteria often use a polysaccharide **capsule**, an extracellular sugar polymer coating surrounding the microbial cell, to surmount host defenses (*1*). Hundreds of structures have been reported from animal and plant pathogens. A very small number of microbial species produce capsular polymers that are chemically identical or similar to acidic **glycosaminoglycans** [GAGs] found in most higher animals (*2*). GAGs are long linear heteropolysaccharides composed of repeating disaccharide units containing a derivative of an amino sugar (either glucosamine or galactosamine). **Hyaluronan** [HA], **chondroitin**, and **N-acetylheparosan** (or **heparosan**) contain glucuronic acid [GlcUA] as the other component of the disaccharide repeat (Table I). Most vertebrate tissues contain these GAGs, but the chondroitin chains and the heparosan chains are further modified after sugar polymerization by sulfation, deacetylation, and/or epimerization; the known microbes do not perform these modification reactions. The bacterial GAG capsules are not immunogenic and assist in avoiding host defenses (*1*,*2*). Due to their chemical similarity to the vertebrate GAGs, the bacterial polymers are also useful or promising materials for current or future biomedical applications.

Three of the most prominent capsular types of ***Pasteurella multocida***, a widespread Gram-negative animal bacterial pathogen, are sources of acidic GAGs. Carter Types A, D, or F produce hyaluronan, heparosan, or chondroitin polysaccharides, respectively (Table I; *2*). This report summarizes our sugar engineering approach employing the *Pasteurella* biosynthetic enzymes.

Table I. *Pasteurella* Glycosaminoglycans and Synthases

Polysaccharide	*Repeat Structure*	*P. multocida*	*enzyme*
HA, Hyaluronan	[β4 GlcUA- β3 GlcNAc]	Type A	pmHAS
Chondroitin	[β4 GlcUA- β3 GalNAc]	Type F	pmCS
Heparosan	[β4 GlcUA- α4 GlcNAc]	Type D	pmHS

Pasteurella Glycosaminoglycan Synthases

Most glycosyltransferase enzymes catalyze the transfer of only one specific type of monosaccharide to an acceptor molecule. In contrast, the various *Pasteurella* GAG glycosyltransferases, called **GAG synthases**, required for the production of the GAG chain transfer two distinct monosaccharides to the growing chain in a repetitive fashion (*2*). In this microbe, as well as all other known organisms, the enzymes that synthesize the alternating sugar repeat backbones utilize **UDP-sugar precursors** and **metal cofactors** (*e.g.* magnesium and/or manganese ion) near neutral pH according to the overall reaction:

$$n\ \textbf{UDP-GlcUA} + n\ \textbf{UDP-HexNAc} \rightarrow \textbf{2}n\ \textbf{UDP} + [\textbf{GlcUA-HexNAc}]_n \quad (1)$$

where HexNAc = GlcNAc (N-acetylglucosamine) or GalNAc (N-acetylgalactosamine). Depending on the specific GAG and the particular microbe examined, the degree of polymerization, *n*, ranges from $\sim 10^{2\text{-}4}$

The native *Pasteurella* and the recombinant *Escherichia coli*-derived preparations of the various microbial *Pasteurella* GAG synthases rapidly form long polymer chains *in vitro* (*2*). The **sugar transfer specificity** of the native-sequence enzymes is exquisite and only the authentic sugars are incorporated into polymer products. For example, the native synthase enzymes do not utilize significantly the C4 epimer precursors in comparison to the natural UDP-sugars.

The native bacterial GAG glycosyltransferase polypeptides are associated with the cell membranes; this localization makes sense with respect to synthesis of polysaccharide molecules destined for the cell surface. The first *Pasteurella* GAG synthase to be identified was the 972-residue HA synthase from Type A strains, **pmHAS** (Table I). This single polypeptide transfers both sugars, GlcNAc and GlcUA, to form the HA disaccharide repeat (*3*).

The chondroitin chain is chemically identical to HA except that GalNAc is substituted for GlcNAc. The 965-residue Type F enzyme, **pmCS**, which has

~90% identity at the gene and protein level to pmHAS, polymerizes the chondroitin chain (*4*).

The 617-residue Type D *Pasteurella* heparosan synthase, **pmHS**, is not very similar at the protein level to either pmHAS or pmCS (*5*). The pmHS enzyme, however, resembles a fusion of the *E. coli* K5 KfiA and KfiC proteins (note that two individual polypeptides are required for polymer formation) responsible for making heparosan in this human pathogen. Even though HA and heparosan polysaccharides have identical sugar compositions, the basic enzymology of their synthases must differ. All UDP-sugar precursors are alpha-linked. HA is an entirely beta-linked polymer, therefore, only inverting mechanisms are utilized during the sugar transfer. On the other hand, heparosan contains alternating alpha- and beta-linkages implying that both a retaining and an inverting mechanism are involved in synthesis. The production of the two types of anomeric glycosidic bonds probably requires distinct catalytic sites.

Elongation Activity of Recombinant *Pasteurella* Synthases

The *E. coli*-derived recombinant pmHAS, pmCS, and pmHS will elongate exogenously supplied GAG-oligosaccharide acceptors *in vitro* (*4-7*). Unexpectedly, the native enzymes from *Pasteurella* do not exhibit this activity at significant levels *in vitro*.

The advent of molecularly cloned GAG synthase genes makes it is possible to prepare virgin enzymes lacking a nascent GAG chain if the proper host is utilized for expression (*e.g.* lacks the required UDP-GlcUA). We speculate that the recombinant enzymes will elongate sugars that enter the accessible acceptor-binding site; normally this site contains an elongating GAG chain *in vivo*. The site of nascent GAG chain binding and/or polymer growth is not yet known for any *Pasteurella* synthase, but it should be proximal or accessible to the UDP-sugar binding sites.

The *Pasteurella* GAG synthases **add sugars to the non-reducing reducing terminus of the linear polymer chain** as determined by testing defined acceptor molecules (*6*). Therefore, the reducing termini of the acceptor sugar may be modified without compromising elongation. We have found that various tagged (*e.g.* radioactive or fluorescent) or immobilized (*e.g.* on plastic or glass surfaces) acceptor molecules are elongated by pmHAS (*6*), pmCS, or pmHS.

Experiments first with recombinant pmHAS demonstrated that **single sugars are added to the growing chain sequentially**; the intrinsic fidelity of each transfer step assures the production of the GAG repeat structure (Fig. 1: *6*). The homologous chondroitin synthase, pmCS, behaves in a similar fashion (*4*). Thus other possible synthetic reactions include:

UDP-GlcUA + [HexNAc-GlcUA]$_n$ → UDP + GlcUA-[HexNAc-GlcUA]$_n$ (2)
UDP-HexNAc + [GlcUA-HexNAc]$_n$ → UDP + HexNAc-[GlcUA-HexNAc]$_n$ (3)

However, as described later, the sugar acceptor specificity is not as precise as the sugar transfer specificity allowing one synthase to extend non-cognate GAG polymers to create chimeric polysaccharides.

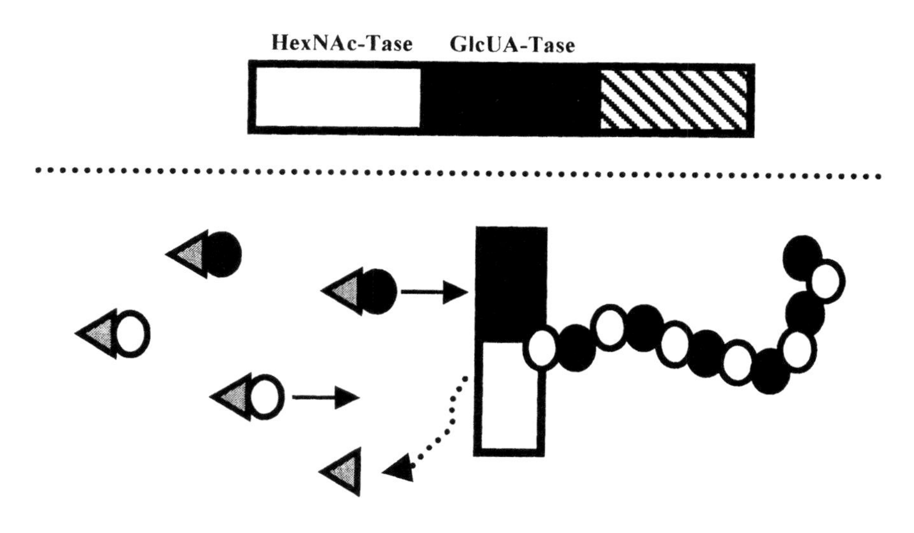

Figure 1. Schematics of pmHAS and pmCS domain structure and the model of the reaction mechanism. Top: Two separate glycosyltransferase (Tase) active sites exist in one polypeptide that add the hexosamine (white) or the glucuronic acid (black). The membrane-associated region (crosshatched) at the carboxyl terminus may be removed to yield a more practical soluble synthase catalyst (8). Bottom: The monosaccharides from the UDP-sugars (UDP, gray triangle; HexNAc, white circle; GlcUA, black circle) are added individually to the non-reducing terminus of the nascent chain in a stepwise fashion. (Reproduced with permission from reference 9. Copyright 2002.)

Domain Structures of *Pasteurella* GAG Synthases

The pmHAS enzyme has been shown to contain **two independent glycosyltransferase sites** by biochemical analysis of various mutants (Fig. 1; *8*). The GlcNAc-transferase or the GlcUA-transferase activities of the *Pasteurella* enzyme can be assayed separately *in vitro* by supplying the appropriate acceptor oligosaccharide and only one of the UDP-sugar precursors (as in reactions 2 or 3). Two tandemly repeated sequence elements are present in pmHAS. Each element of pmHAS contains a short sequence motif containing Aspartate-Glycine-Serine. Mutation of the aspartate residue in any one motif of pmHAS converts the dual-action synthase into a single-action glycosyltransferase (*8*).

The *Pasteurella* chondroitin synthase, pmCS, contains separate GalNAc-transferase (a slightly mutated version of the GlcNAc-site of pmHAS) and GlcUA-transferase sites. The pmHS domains are still be investigated. These single-action enzymes are useful catalysts for the preparation of defined oligosaccharides described later.

Hybrid GAG Polymers

We recently found that the various *Pasteurella* GAG synthases can also use **non-cognate acceptor molecules**. In the example shown in Fig. 2, we added a HA chain onto an existing chondroitin sulfate chain using pmHAS (*9*). This particular family of hybrid molecules may be useful as artificial proteoglycans for tissue engineering that do not contain the protein component of the natural cartilage proteoglycans.

The full repertoire of hybrid GAG molecules possible by this technique is vast. The variables of *i*) the identity of the GAG components, *ii*) the number (*e.g.* 2, 3, or more GAG types) and *iii*) the placement (*e.g.* relative orientation with respect to the reducing terminus) of sugar components in one chain, and *iv*) the size of each segment (*e.g.* large or short stretches of a given polymer) appear to be very flexible (*9*).

Synthesis of Defined, Monodisperse Oligosaccharides

We have developed the chemoenzymatic synthesis of monodisperse GAG oligosaccharides (*10*). Potential medical applications for HA oligosaccharides (n = ~3-10) include killing cancerous tumors and enhancing wound vascularization. The *Pasteurella* HA synthase, a polymerizing enzyme that

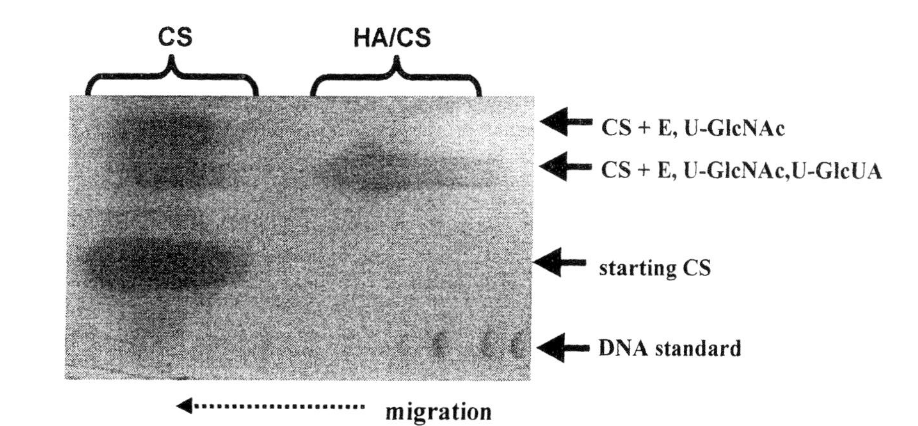

Figure 2. Agarose gel analysis of hybrid HA/Chondroitin sulfate polymer. Chondroitin sulfate (CS) was extended by reaction in vitro with pmHAS (E) and both UDP-sugars to form larger polymers with HA extensions. In the control reaction with chondroitin and only one UDP sugar, no hybrid polymer is observed.

normally elongates HA chains rapidly (~1 to 100 sugars/second) as in reaction 1, was converted by mutagenesis into two single-action glycosyltransferases. The resulting GlcUA-transferase and GlcNAc-transferase are appropriate for performing reactions 2 or 3, respectively. For convenience, soluble forms of mutant pmHAS truncated at the carboxyl termini (*8*) were purified and immobilized onto beads for utilization as solid-phase catalysts (Fig. 3). Two **immobilized enzyme-reactors** were used in an alternating fashion to produce quantitatively desirable sugars in a controlled, stepwise fashion without purification of the intermediates. This technology platform is also amenable to the synthesis of tagged (fluorescent-, medicant- or radioactive-labeled) oligosaccharides for biomedical testing (Fig. 4).

Construction of Oligosaccharide Libraries

The reactions 2 and 3 used in the oligosaccharide synthesis may also be performed on acceptors immobilized on a solid phase (*7*). We are in the process of constructing libraries of unique small GAG oligosaccharides. The libraries are made by first chemically linking an appropriate acceptor (*e.g.* amino-HA4) to spots on a glass slide in an array format or to plastic wells of a microtiter plate (Fig. 5). The acceptor is then elongated by a single sugar with a synthase. Limiting the UDP-sugar that is available with the chosen enzyme can control the reaction at each step in each spot or well. For example, various combinations of substrate and the appropriate enzyme (*e.g.* UDP-GlcNAc + pmHAS, or UDP-GalNAc + pmCS, or UDP-GlcUA + pmHAS) are used in an alternating fashion at each spot or well. Many unique molecules will be made in parallel. Each spot or well has a unique, known structure because the reaction sequence is defined. These sugar-based arrays will be then be tested for binding to protein targets implicated in disease (Fig. 6). We expect to find new and novel lead compounds for further study in cultured cell and animal models.

Novel GAGs with Chimeric Synthases

Typically, the native sequence GAG synthases will only transfer the authentic UDP-sugars to produce the natural polymer. However, we have now created new enzymes that can make novel polymers that are not known to exist in Nature. We used a chimeric genetic modification approach to map out the Gal/Glc specificity for UDP-hexosamines of the *Pasteurella* pmHAS [HA

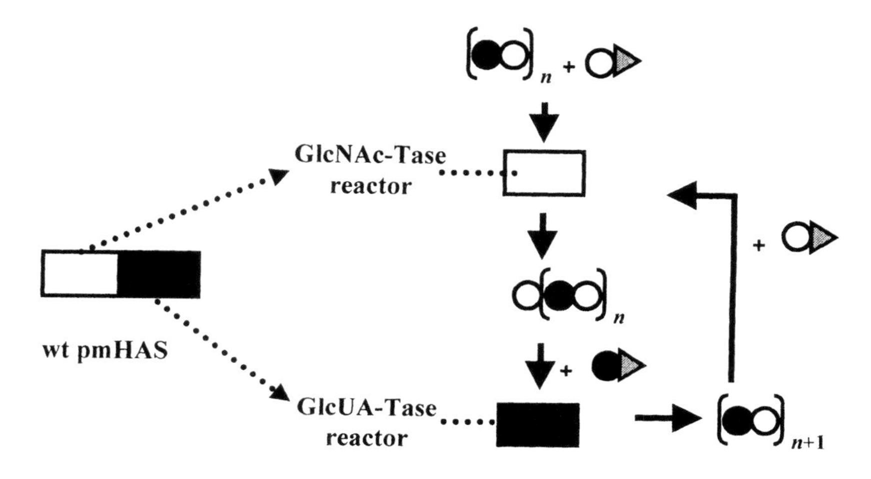

Figure 3. Schematic of creation and use of catalysts for oligosaccharide synthesis. Site-directed mutagenesis was used to produce two single-action enzymes from wild-type pmHAS. Immobilized enzyme beads add one sugar at a time to form short HA oligosaccharides in a defined stepwise fashion. (Reproduced with permission from reference 10. Copyright 2002.)

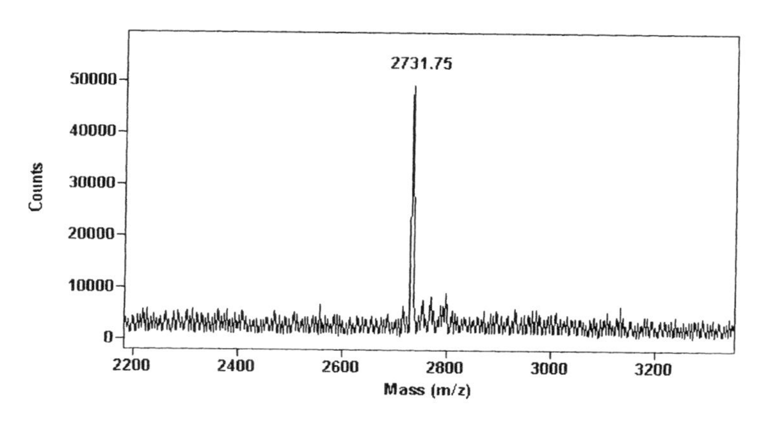

Figure 4. MALDI-TOF MS analysis of fluorescent-HA12. The strategy shown in Fig. 3 was utilized to elongate a tagged HA4 molecule into a monodisperse HA12 molecule. No purification during the 8-step procedure was required. (Reproduced with permission from reference 10. Copyright 2002.)

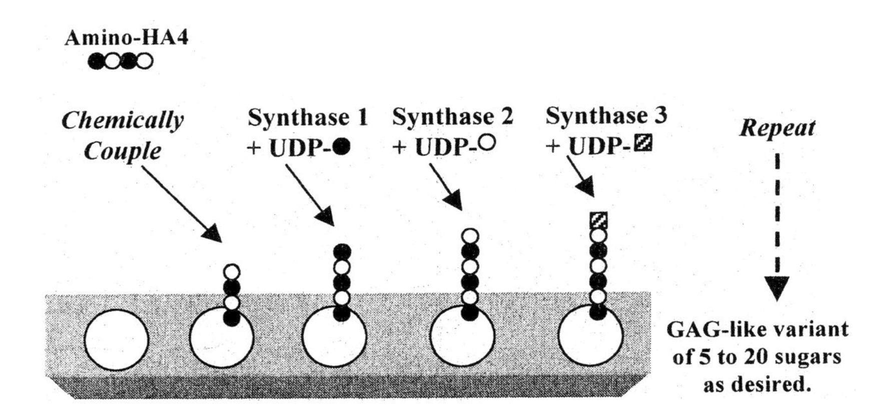

Figure 5. Schematic of synthesis of a variant GAG sugar in a spot or well of an immobilized library. The construction of a single, known defined oligosaccharide is illustrated. A series of spots or wells each containing one species of a wide variety of molecules may be generated in parallel by using a different sequence of synthase/UDP-sugar combinations for each spot or well. (Reproduced with permission from reference 10. Copyright 2002.)

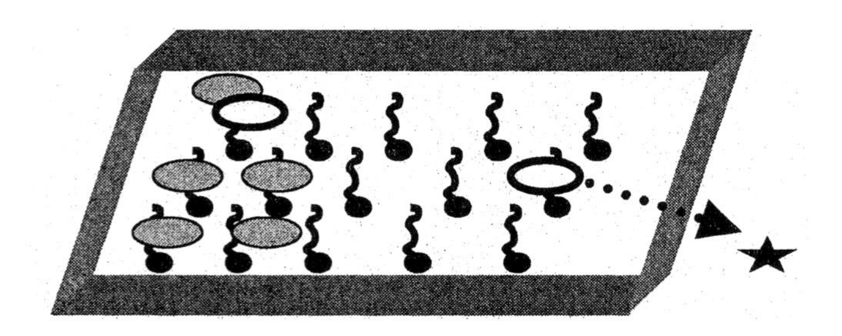

Figure 6. Screening of a GAG oligosaccharide library. A slide or a microtiter plate with a wide variety of distinct sugars (black squiggle) is screened with various GAG-binding proteins (gray or white ovals). Any oligosaccharide with desirable binding characteristics (star - e.g. high affinity and/or selectivity) may be produced in larger amounts utilizing the immobilized enzyme-reactors.

synthase] and pmCS [chondroitin synthase] enzymes (Fig. 7; *11*). One of the new chimeric enzymes, pmCHC, composed of several regions of pmHAS/pmCS together in one polypeptide will **add either GlcNAc or GalNAc** based on the UDP-sugar availability (Table II). The new enzyme has relaxed specificity at the C-4 position of the substrate. For example, if one adds a mixture of UDP-GlcNAc and UDP-GalNAc to the reaction mixture, then the new synthase will make a novel blended hyaluronan/chondroitin co-polymer with potential for different activities. The blended co-polymers may have improved properties as a hyaluronan or a chondroitin substitute in therapeutic treatment including joint and eye supplementation or prevention of post-surgical adhesions.

We also found that an **unnatural precursor**, UDP-glucosamine, is also incorporated by the pmHAS/pmCS chimera (Table II; *11*). The resulting sugar polymer will contain free amino groups; these functional groups should be very useful for coupling medicants to the polymer or for crosslinking polymers to form gels (Fig. 8).

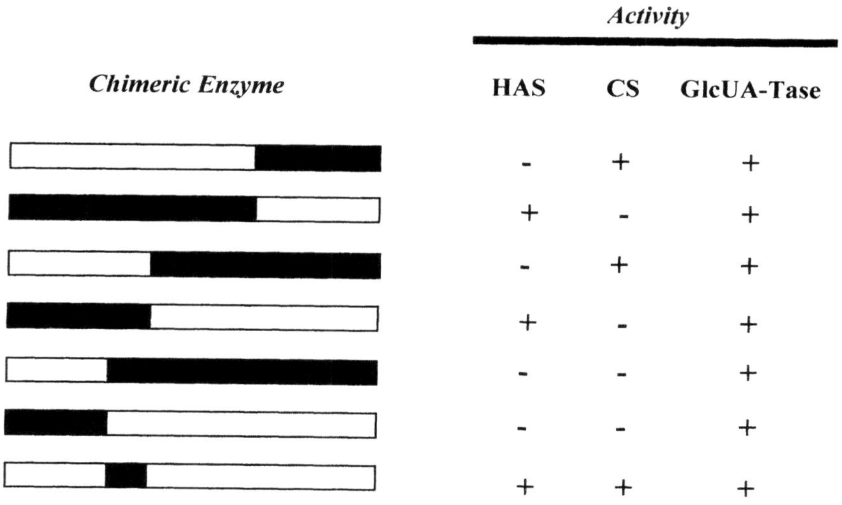

Chimeric Enzyme	*Activity*		
	HAS	**CS**	**GlcUA-Tase**
	-	+	+
	+	-	+
	-	+	+
	+	-	+
	-	-	+
	-	-	+
	+	+	+

Figure 7. Analysis of various chimeric synthases. Portions of the pmHAS (black) and pmCS (white) enzymes were fused to produce chimeric enzymes. The constructs were tested for sugar transfer activity (HAS, HA synthase; CS, chondroitin synthase; GlcUA transferase) in vitro. The last enzyme, pmCHC, had relaxed specificity and could incorporate either UDP-hexosamine.

Table II. Sugar Specificity of pmCHC

substrate sugar	*relative incorporation*
UDP-GalNAc	100%
UDP-GlcNAc	28%
UDP-Glucosamine	2%
UDP-Galactose	not detectable
UDP-Glucose	not detectable
UDP-Mannose	not detectable
UDP-Xylose	not detectable

NOTE: A series of reactions containing pmCHC + UDP-[^{14}C]GlcUA + various UDP-sugar substrates was used to assess sugar specificity of the chimeric synthase. The signal in the assay corresponds to incorporation of radiolabel into high molecular weight polymer. The incorporation in the reaction with UDP-GalNAc was set to 100%.

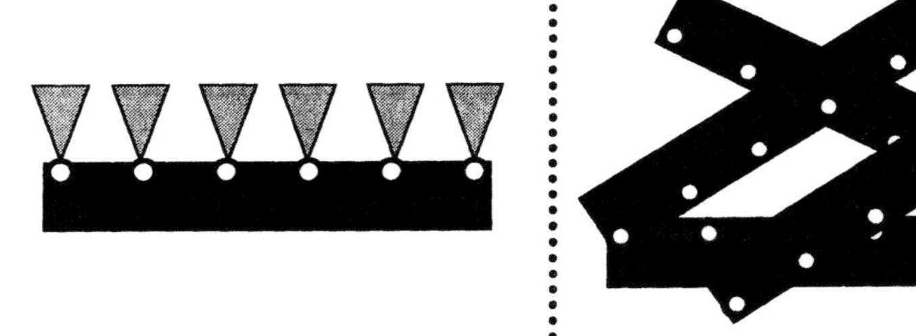

Figure 8. Diagrams of potential biomedical agents using novel amino-GAGs. The amino-containing polymer (black bar) produced using pmCHC and UDP-glucosamine may be derivatized with a medicant (left; gray triangle, drug) or crosslinked to form a gel (right) due to the presence of new useful functional groups (white circles).

Conclusions

We have identified and harnessed useful catalysts for the creation of a variety of defined GAG or GAG-like polymers. A wide spectrum of potential biomedical products for use in the areas of cancer, coagulation, infection, tissue engineering, drug delivery, surgery, and viscoelastic supplementation are possible. We discovered these versatile, malleable *Pasteurella* synthases between 1998 to 2002 thus we have just scratched the surface of the mountain of possibilities for sugar engineering biotechnology.

Acknowledgements: The work described was supported in part by grants from the National Science Foundation (MCB-9876193), the National Institutes of Health (R01-GM56497), and Hyalose, L.L.C.

References

1. Roberts, I.S. *Annu. Rev. Microbiol.* **1996**, *50*, 285-315.
2. DeAngelis, P.L. *Glycobiology* **2002**,*12*, 9R-16R.
3. DeAngelis, P.L.; Jing, W.; Drake, R.R.; Achyuthan, A.M. *J. Biol. Chem.* **1998**, *273*, 8454-8458.
4. DeAngelis, P.L.; Padgett-McCue, A.J. *J. Biol. Chem.* **2000**, *275*, 24124-24129.
5. DeAngelis, P.L.; White, C.L. *J. Biol. Chem.* **2002,** *277*, 7209-7213.
6. DeAngelis, P.L. *J. Biol. Chem.* **1999**, *274*, 26557-26562.
7. DeAngelis, P.L. U.S. Patent 6,444,447, 2002.
8. Jing, W.; DeAngelis, P.L. *Glycobiology*. **2000**, *10*, 883-889.
9. DeAngelis, P.L. unpublished.
10. DeAngelis, P.L.; Oatman, L.C.; Gay, D.F. unpublished.
11. Jing, W; DeAngelis, P.L. unpublished.

Chapter 10

Production of Oligosaccharides Using Engineered Bacteria: Engineering of Exopolysaccharides from Lactic Acid Bacteria

Laure Jolly, Véronique Tornare, and Sunil Kochhar

Nestlé Research Center, Vers-chez-les-Blanc, P.O, Box 44, 1000 Lausanne 26, Switzerland

To date, there is no feasible way to produce oligosaccharides in high amount in food grade conditions at competitive prices. Rapid advances in the cloning and expression of glycosyltransferase genes, especially from bacteria, could open the way to overcoming difficulties in the mass production of oligosaccharides. The large-scale production of oligosaccharides using either glycosyltransferases isolated from engineered microorganisms or whole cells as an enzyme source could promote a new era in the field of carbohydrate synthesis.

Carbohydrates play important roles in numerous biological processes (e.g., as carbon source, signal mediator, attachment mediator). In the gastrointestinal tract, a complex microbial system is established starting at birth and continuing throughout life. Wishfully, a beneficial and commensal flora should predominate at all times. One strategy is to design specific multivalent carbohydrate carriers to promote a beneficial flora and to reduce harmful microbes.

Synthesis of oligosaccharides is cumbersome and purification difficult and time consuming. This very challenging field is moving fast. The last few years, a couple of new chemical or chemo-enzymatic methods have been described in the literature. Technologies to chemically synthesize these oligosaccharides have dramatically advanced mainly due to the introduction of good anomeric leaving groups. Solution-phase methodologies, the status of solid-phase synthesis of oligosaccharides, and combinatorial synthesis of oligosaccharide libraries have been reviewed by Kanemitsu et al. (*1*). However, large-scale chemical production of oligosaccharides remains expensive and prohibitive for industrial application as nutritional ingredients. Among the chemo-enzymatic synthesis of oligosaccharides, covalently immobilizing recombinant glycosyltransferases (GTFs) with activated Sepharose beads has been employed for the practical synthesis of trisaccharide derivatives (*2*), or homogeneously soluble PEG polymer with a multi-enzyme system has been shown to allow the production of the Galili epitope, which acts against the Toxin A from *Clostridium difficile* (*3*). Although these methods showed promising results, the use of non food-grade materials prevent them from applications in the food industry. The human milk oligosaccharide (HMO) lacto-*N*-neotetraose (LNnT) has been successfully synthesized by using a chemo-enzymatic approach employing a galactosidase rather than a GTF (*4*). However, the lack of stereoselectivity of transglycosylation reaction is often a bottleneck for efficient synthesis. Most promising approaches are in vivo production of oligosaccharides using metabolically engineered bacteria. HMOs could successfully be obtained in high yields (*5*, *6*, *7*, *8*). These methods are based on the utilization of whole cells that overexpress genes coding for GTF that are naturally involved in the synthesis of oligosaccharides from sugar-nucleotides and on systems that allow a sugar-nucleotide recycling by these whole cells. Using food grade bacteria such as lactic acid bacteria (LAB) is the main issue for the food industry.

In recent years, exopolysaccharides (EPSs) produced by LAB have attracted much attention in the food industry because of their rheological properties combined with their GRAS (generally regarded as safe) status (*9*). Certain EPSs produced by LAB are also claimed to have beneficial physiological effects. It is speculated that the increased viscosity of EPS containing foods may increase the residence time of ingested fermented milk in the gastrointestinal tract and, coupled to a low degradability of EPSs (*10*), might therefore be beneficial to a transient colonization by probiotic bacteria (*11*). A further example of a suggested health benefit of some EPSs is the generation of short-chain fatty acids (SCFAs) upon degradation in the gut by the colonic microflora. SCFAs provide energy to epithelial cells and some have been claimed to play a role in the prevention of colon cancer (*12*, *13*). *In vivo* studies by oral administration of EPSs will be crucial to demonstrate clearly the different health beneficial properties that have been mentioned in the literature so far (e.g., antitumor, antiulcer, immunomodulating or cholesterol-lowering activities).

However, one way to design desired biologically active carbohydrates could be to modify the natural EPS structures (*9*, *14*). Recently, Vincent et al. (*15*) determined the structure of a polysaccharide produced by *S. macedonicus* Sc136 that contains the trisaccharide sequence β-D-Glc*p*NAc-(1→3)-β-D-Gal*p*-(1→4)-β-D-Glc*p*. This corresponds to an internal EPS backbone of lacto-*N*-tetraose and LNnT. In fact, the same trioses have been identified in the structure of several human milk oligosaccharides that are important for infant nutrition. Deleting and inserting new genes coding for GTFs can be one way to engineer functional EPSs, based on an existing template such as the EPS produced by *S. macedonicus* Sc136, similarly to the approach based on LPS in *Escherichia coli* that led to *in vivo* neutralization of the Shiga toxin in mice (*16*).

In vivo modification of an EPS

Our strategy was to use EPS in LAB to design specific multivalent carbohydrate carriers to promote a beneficial flora and to reduce harmful microbes. Such designed probiotic LAB will allow the delivery of functional carbohydrates in the gastrointestinal tract at the site where needed. As a prototype system, *S. macedonicus* Sc136' EPS and malignant microbes have been chosen. The trisaccharide sequence β-D-Glc*p*NAc-(1→3)-β-D-Gal*p*-(1→4)-β-D-Glc*p*, present in the EPS repeating unit (Figure 1) (*15*), corresponds to the internal backbone of the lacto-*N*-tetraose and LNnT units, that are the structural core of a majority of HMOs. This property offers the potential to engineer EPS to neutralize pathogens such as enteropathogenic *E. coli* (EPEC) or *Clostridium difficile* toxin A. The *eps* gene cluster from *S. macedonicus* Sc136 has been elucidated. It contains genes coding for GTFs required for the biosynthesis of the repeating unit. These enzymes have been purified and their sugar specificity identified. Knockout mutants of genes coding for relevant GTFs and heterologous overexpression of GTFs have been performed in order to modify the EPS side chain. A second approach consisted in extracting the EPS first and then modifying it in vitro, in order to use them as directly as ingredients. As a prototype system, *Lactobacillus delbrueckii bulgaricus* Lb295 and similar malignant microbes have been chosen. The disaccharide β-D-Gal*p*-(1→4)-β-D-Glc*p* (lactose motif), present in the side chain of the repeating unit can be fucosylated to display a disaccharide α-(1,2)-D-Fuc-β-D-Gal*p*-(1→4)-β-D-Glc*p* motif, to mimic the abundant 2'-FL HMO. A α-(1,2) -fucosyltransferase from *Helicobacter pylori* has been cloned, overexpressed, purified, and results on EPS fucosylation will be presented.

→4)-α-D-Glc*p*-(1→4)-β-D-Gal*p*-(1→4)-β-D-Glc*p*-(1→
3
↑
1
β-D-Gal*f*-(1→6)-β-D-Glc*p*-(1→6)-β-D-Glc*p*NAc

Figure 1: Structure of the EPS from *S. macedonicus* Sc136

Eps gene cluster identification

In order to modify in vivo the biosynthesis pathway of EPS, genetic data is a prerequisite. The overall structure of the *eps* cluster of *S. macedonicus* Sc136 (Figure 2) was found to be similar to other ones described in LAB (*17*). The sequence of the functions of genes is as follows: regulation, chain length determination, biosynthesis of the repeating unit, polymerization and export. More precisely, the oligosaccharide repeating unit is first assembled by the sequential transfer of sugar residues onto a lipophilic carrier by specific GTFs. Unlike the other GTFs, the first GTF does not catalyze a glycosidic linkage but transfers a sugar-1-phosphate onto a lipophilic anchor, like undecaprenylphosphate. Subsequently, the completed repeating unit is exported, polymerized and released in the case of secreted polysaccharides.

GTFs: biosynthetic tools

The biosynthesis of EPS involves the build-up of individual repeating units on a lipid carrier. The GTFs are key enzymes for this biosynthesis of the repeating unit, since they catalyze the transfer of sugar moieties from activated donor molecules to specific acceptor molecules, thereby forming a glycosidic bond (*17*). So far, linear alignment methods have shown that GTFs share local amino acid sequence similarities, but low level of similarity often encountered among these enzymes sometimes prevents the alignment of related sequences. For these low level similarities, hydrophobic cluster analysis (HCA), which is a powerful sequence comparison method able to detect three-dimensional similarities in proteins with very limited sequence relatedness, was applied. The large variety of GTFs can be classified according to the regio and the stereochemistry. α- and β-GTFs (retaining or inverting GTFs) have different catalytic mechanisms and it was possible to distinguish between them by similarity searches and sequence comparison (*18*).

So far, very few GTFs from LAB have been fully characterized. Only recently some were heterologously expressed and purified and their substrates characterized through the use of a synthesized acceptor mimicking the natural substrate (*19*, *20*). This allowed for the first time direct determination of GTF specificities. Furthermore, an example of *in vitro* polyprenol-linked oligosaccharide synthesis based on purified GTFs was shown, with a trisaccharide product. This emphasized that bacterial GTFs present an alternative to chemical synthesis for oligosaccharide production. Although the major advantage is the high stereospecificity of the reaction, one limitation is the number of available enzymes. The next major challenge in the study of GTFs might be to get a better understanding of their mode of action of GTFs in order to modulate their specificity by mutagenesis manipulation. So far, even though X-

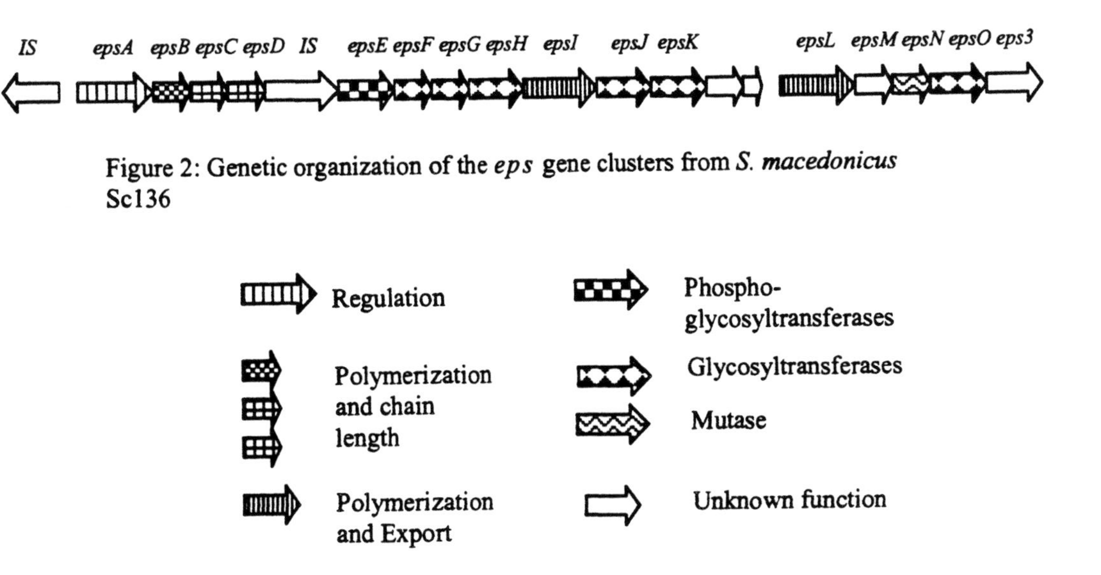

Figure 2: Genetic organization of the *eps* gene clusters from *S. macedonicus* Sc136

ray structures allow some preliminary assumptions, the determination of new three-dimensional structures is an essential requirement to tackle this issue (*21, 22*). Furthermore, although one structure succeeded in cocrystallizing acceptor and / or donor, basis of acceptor and donor binding specificities could not fully be elucidated (*23*).

With LAB and EPS, we have now in hand a dozen of GTFs with available acceptors and donors and different stereochemistry and regio specificities (*20*). In *S. macedonicus* Sc136, gene products sharing homologies with GTFs (*epsFG, epsH, epsJ* and *epsK*) have been studied (*unpublished*). Hydrophobic cluster analyses has allowed assigning putative functions. In order to validate the predictions, the gene products were heterologously expressed, purified and their substrates characterized through the use of the same protocol as described by Jolly et al. (*20*) (Figure 3). EpsF coupled with EpsG codes for a β-(1,4)-galactosyltransferase. EpsH encodes a α-glucosyltransferase with most probably an α-(1,4) stereospecificity and EpsJ, a β-*N*-acetylglucosamine transferase with most probably a β-(1,3) stereospecificity. Function of EpsK remains unclear. According to the EPS structure, two GTFs activities were missing: the β-(1,6)-glucosyltransferase activity and the β-(1,6)-galactofuranosyltransferase activity. Either EpsK carries out the two activities as the two linkages formed are identical or EpsK carries out only one of the activity and one gene coding for a GTF has not been identified (Figure 4).

In vivo EPS engineering

The objective here was to modify the EPS produced by *S. macedonicus* Sc136 in order to display a LNnT motif on the surface of the bacteria. The HMO LNnT has actually been shown to neutralize the enteropathogenic *E. coli* pathogen (*24*). In order to display this motif on the EPS from *S. macedonicus* Sc136, inhibition of the enzyme coding for the glucosyltransferase adding the glucose on the side-chain and addition of a β-(1,4)-galactosyltransferase instead are required. The first step consists in knocking out *epsK*, the putative gene coding for the glucosyltransferase and then to express the *LgtB* gene from *N. meningitidis* the only available bacterial β-(1,4)-galactosyltransferase able to add a galactose on a GlcNAc (*25*). Knocking out genes in LAB is a first crucial step. A similar approach than the one used for studies on capsular polysaccharides by Cieslewicz et al. was applied (*26*). A non-polar mutant could not be obtained with *epsK* but only with *epsJ*, whose gene product adds the branching sugar on the backbone. The mutated strain was fermented in a defined MRS medium and the EPS extracted and purified. The monosaccharide composition of this EPS was obtained after acid hydrolysis on an HPAE-PAD equipment. Most surprisingly, the ratio of Glc:Gal:GlcNAc was similar to the wild type EPS ratio: 3:2:1. NMR analysis showed that the structure was identical to the wild type EPS structure

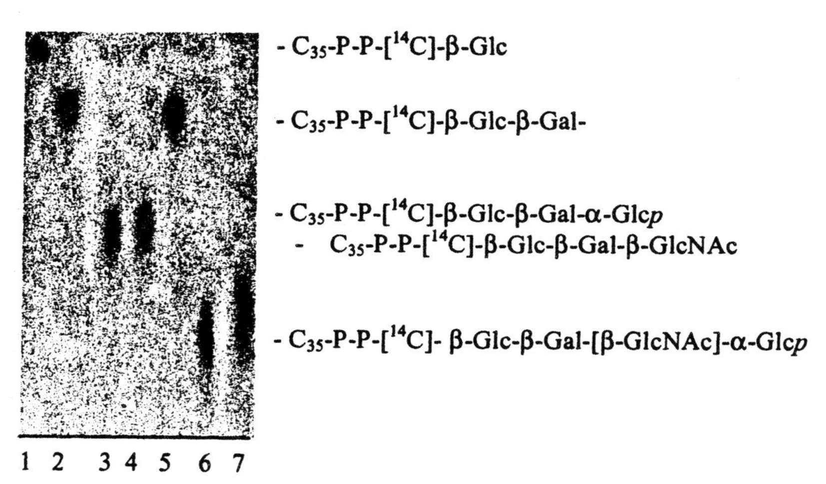

Figure 3: TLC analysis of activity assays with purified GTFs, labeled C_{35}-P-P-β-[^{14}C]Glc and UDP donors, overnight at room temperature. Lane 1: reaction with purified EpsF-EpsG and UDP-Gal. Lanes 2: reaction with purified EpsF-EpsG and UDP-Gal. Lane 3: reaction with both purified EpsF-EpsG and EpsH, UDP-Gal and UDP-Glc. Lane 4: reaction with both purified EpsF-EpsG and EpsJ, UDP-Gal and UDP-GlcNAc. Lane 5: reaction with both purified EpsF-EpsG and EpsK, UDP-Gal, UDP-Glc and UDP-GlcNAc. Lane 6: reaction with purified EpsF-EpsG, EpsH and EpsJ, UDP-Gal, UDP-Glc and UDP-GlcNAc. Lane 7: reaction with both purified EpsF-EpsG, EpsH, EpsJ and EpsK, UDP-Gal, UDP-Glc and UDP-GlcNAc.

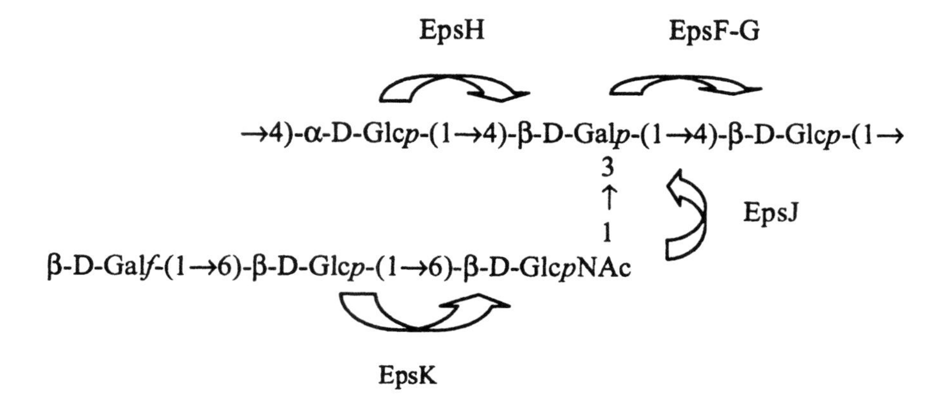

Figure 4: Putative functions of *GTFs* from *S. macedonicus* Sc136. EpsF-G has been fully characterized has a β-(1,4)-galactosyltransferase (data not shown). For EpsH and EpsJ, only the substrate specificity could be determined (Figure 3). According to the EPS structure, a putative function of EpsK could be assigned but could be determined in vitro.

(data not shown). Although it has been checked that in the mutated strain translation of *epsJ* flanking genes were not disturbed by the gene knock-out (data not shown), the wild EPS was still produced. This EPS could not come from another locus, as no EPS production could be detected when the cluster was inactivated (pG-host9 chromosomal insertion). The mutated strain inoculum was not contaminated by the wild type strain, as checked by PCR. One remaining hypothesis was the compensation of the *epsJ* deletion by another gene of *S. macedonicus Sc136*. Among the *eps* genes from the *eps* cluster, *epsO,* whose gene product shares 57% identity with *cpsY* from *Streptococcus salivarius* that codes for a putative hexose hydrolase, was the only potential gene that might have a gene product carrying a sugar-binding domain hence displaying a GTF activity. The gene product was heterologously expressed, purified and checked for GTF activity and in particular a β-(1,3)-*N*-acetylglucosamine transferase activity like EpsJ. EpsO was able to carry out an *N*-acetylglucosamine transferase activity (data not shown). *EpsO* is a potential candidate for complementation of *epsJ* deletion in *S. macedonicus* Sc136.

Knocking-out a gene coding for a GTF to modify the EPS structure is a major prerequisite and seems to be most difficult to achieve. Similar results were obtained with *epsF* in *S. thermophilus* Sfi6: when knocked out, the EPS produced was also similar to the wild type one (*unpublished*). In order to circumvent this hurdle, attempts to overexpress the β-(1,4)-galactosyltransferase LgtB in *S. macedonicus* to compete with EpsK in vivo and obtain an EPS with the desired galactose on the *N*-acetylglucosamine moiety were performed. When *lgtB* could successfully be introduced in *S. macedonicus* Sc136, the EPS produced by this recombinant strain was extracted and analyzed. However, GC-MS analysis could not detect any β-galactose linked to an *N*-acetylglucosamine (data not shown). In summary, an alternative to engineer an EPS could be to introduce a designed *eps* gene cluster in a *L. lactis* non-EPS producing strain. The *eps* cluster from *S. thermophilus* Sfi6 has successfully been integrated in *L. lactis* MG1363 and the EPS produced was similar to the wild type one (*27*). However, these results suggest that more studies are needed to better understand the overall EPS biosynthesis in order to modulate it further.

In vitro modification of an EPS

To engineer EPS in order to display biofunctional motifs, a different approach has been envisaged: modify in vitro the structure of a purified EPS. However, very few EPSs carrying the useful side-chain motif are available. Some attempts have been performed to produce an EPS that displays the HMO 2-fucosyllactose motif (2'-FL), a very abundant oligosaccharide in human milk (*24*). According to the literature, the only EPS that carries side chain lactose amenable to fucosylation was the *L. delbrueckii bulgaricus* Lb291 EPS (*28*, *29*).

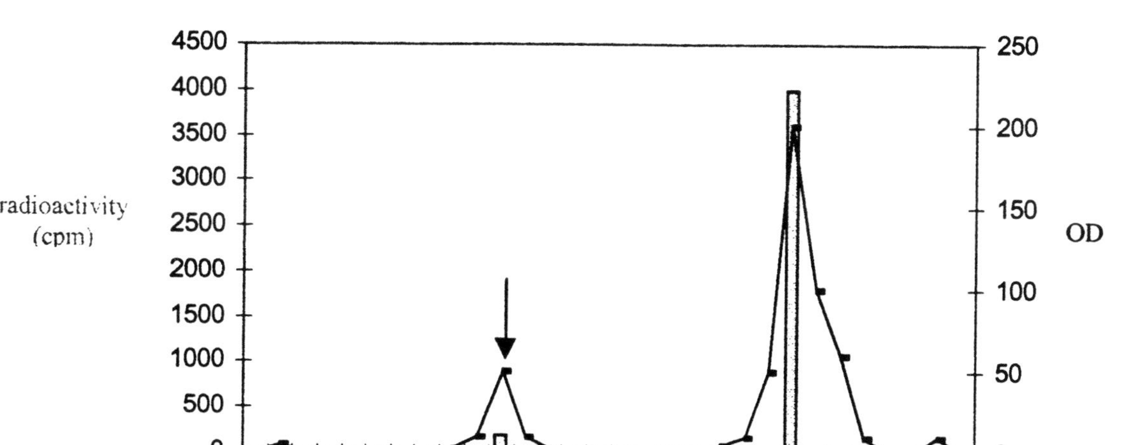

Figure 5: Fucosylation of the EPS from *Lb. delbrueckii bulgaricus* Lb291. Size exclusion chromatography of the reaction mixture containing EPS, GDP-fucose, GDP- [^{14}C]-fucose, $MnCl_2$ and purified α-(1,2)-fucosyltransferase from *H. pylori*. Level of radioactivity (cpm) has been determined by using a scintillation counter. The optical density was detected at 280 nm.

Attempts of fucosylation have been carried by using a α-(1,2)-fucosyltransferase from *H. pylori*. The heterologously expressed and purified enzyme was able to transfer a fucose on lactose acceptors like C_{35}-P-P-β-Glc-(1←4)-β-[^{14}C]Gal (data not shown). When tested with the purified EPS as an acceptor and labeled GDP-[^{14}C]-fucose as a donor, the fucosylation levels were very low whatever the ratio of acceptor / donor used. The EPS fraction free of GDP-fucose was collected by size exclusion chromatography and its radioactivity counts determined (Figure 5). However, the incorporation level of [^{14}C]-fucose was low. Nevertheless, the data was highly reproducible and revealed that 1% of EPS was fucosylated. The yield needs to be improved for any application. One alternative could be to partially depolymerize the polysaccharide that could lead to a better acceptor for the GTF and enhance the reaction rate. However, depolymerization is difficult to control. Smith degradation can not be carried out on this EPS, as no β(1,3) linkage exists. By using TFA at different temperature and different time, some oligosaccharides could be obtained, but either most of the EPS was not depolymerized or was completely depolymerized to mono- and disaccharides.

Conclusions

EPS engineering has been partially successful. Yields of in vitro glycosylation by using bacterial GTFs need to be improved for any further industrial application. The EPS from *Lb. bulgaricus delbrueckii* Lb291 has been partially fucosylated, leading to an EPS most probably carrying a 2'-FL HMO motif on the side chain. To genetically modify genes coding for GTFs in LAB to modulate the structure of an EPS requires further studies. To circumvent compensation issue when knocking out genes, one approach could be to introduce an *eps* cluster carrying the desired GTFs in a non-EPS producing *L. lactis* strain. More fundamental studies on the basic understanding on the overall biosynthesis of an EPS are required.

References

1. Kanemitsu, T.; Kanie, O. *Comb. Chem High Throughput. Screen.* **2002**, *5*, 339-360.

2. Nishiguchi, S.; Yamada, K.; Fuji, Y.; Shibatani, S.; Toda, A.; Nishimura, S. *Chem Commun (Camb.)* **2001**, 1944-1945.

3. Brinkmann, N.; Malissard, M.; Ramuz, M.; Romer, U.; Schumacher, T.; Berger, E. G.; Elling, L.; Wandrey, C.; Liese, A. *Bioorg. Med Chem Lett* **2001,** *11*, 2503-2506.

4. Murata, T.; Inukai, T.; Suzuki, M.; Yamagishi, M.; Usui, A. T. *Glycoconj. J* **1999,** *16*, 189-195.

5. Blixt, O.; Brown, J.; Schur, M. J.; Wakarchuk, W.; Paulson, J. C. *J Org. Chem* **2001,** *66*, 2442-2448.

6. Dumon, C.; Priem, B.; Martin, S. L.; Heyraud, A.; Bosso, C.; Samain, E. *Glycoconj. J* **2001,** *18*, 465-474.

7. Endo, T.; Koizumi, S. *Curr. Opin. Struct. Biol.* **2000,** *10*, 536-541.

8. Priem, B.; Gilbert, M.; Wakarchuk, W. W.; Heyraud, A.; Samain, E. *Glycobiology* **2002,** *12*, 235-240.

9. Jolly, L.; Vincent, S. J.; Duboc, P.; Neeser, J. R. *Antonie van Leeuwenhoek* **2002,** *82*, 367-374.

10. Ruijssenaars, H. J.; Stingele, F.; Hartmans, S. *Curr. Microbiol.* **2000,** *40*, 194-199.

11. German, B.; Schiffrin, E. J.; Reniero, R.; Mollet, B.; Pfeifer, A.; Neeser, J. R. *Trends Biotechnol.* **1999,** *17*, 492-499.

12. Harris, P. J.; Ferguson, L. R. *Mutation Research* **1993,** *290*, 97-110.

13. Cummings, J. H.; Englyst, H. N. *Am. J Clin Nutr.* **1995,** *61*, 938-945.

14. van Kranenburg, R.; Boels, I. C.; Kleerebezem, M.; de Vos, W. M. *Curr. Opin. Biotechnol.* **1999,** *10*, 498-504.

15. Vincent, S. J.; Faber, E. J.; Neeser, J. R.; Stingele, F.; Kamerling, J. P. *Glycobiology* **2001,** *11*, 131-139.

16. Paton, A. W.; Morona, R.; Paton, J. C. *Nat. Med.* **2000,** *6*, 265-270.

17. Jolly, L.; Stingele, F. *Int. Dairy J.* **2001,** *11*, 733-745.

18. Campbell, J. A.; Davies, G. J.; Bulone, V.; Henrissat, B. *Biochem. J.* **1998**, *329*, 719-723.

19. Lamothe, G. T.; Jolly, L.; Mollet, B.; Stingele, F. *Arch. Microbiol.* **2002**, *178*, 218-228.

20. Jolly, L.; Newell, J.; Porcelli, I.; Vincent, S. J.; Stingele, F. *Glycobiology* **2002**, *12*, 319-327.

21. Breton, C.; Mucha, J.; Jeanneau, C. *Biochimie* **2001**, *83*, 713-718.

22. Davies, G. J. *Nat. Struct. Biol.* **2001**, *8*, 98-100.

23. Persson, K.; Ly, H. D.; Dieckelmann, M.; Wakarchuk, W. W.; Withers, S. G.; Strynadka, N. C. *Nat. Struct. Biol.* **2001**, *8*, 166-175.

24. Sprenger, N.; Kusy N.; Grigorov M. *submitted to publication* **2003**.

25. Wakarchuk, W.; Martin, A.; Jennings, M. P.; Moxon, E. R.; Richards, J. C. *J Biol. Chem* **1996**, *271*, 19166-19173.

26. Cieslewicz, M. J.; Kasper, D. L.; Wang, Y.; Wessels, M. R. *J Biol. Chem* **2001**, *276*, 139-146.

27. Germond, J. E.; Delley, M.; D'Amico, N.; Vincent, S. J. *Eur. J Biochem* **2001**, *268*, 5149-5156.

28. De Vuyst, L.; De Vin, F.; Vaningelgem, F.; and Degeest B. *Int. Dairy J.* **2001**, *11*, **687-707**.

29. Faber, E. J.; Kamerling, J. P.; Vliegenthart, J. F. *Carbohydr. Res* **2001**, *331*, 183-194.

Chapter 11

Production of Oligosaccharides by Coupling Engineered Bacteria

Satoshi Koizumi

Tokyo Research Laboratories, Kyowa Hakko Kogyo Company, Ltd., 3–6–6 Asahimachi, Machida, Tokyo 194–8533, Japan

Oligosaccharides on the cell surface have been expected to be potential pharmaceuticals and neutraceuticals, however it has been very difficult to synthesize these oligosaccharides. Recently, identification of the genes of glycosyltransferases in bacteria and expression of the genes in *Escherichia coli* enabled the mass production of glycosyltransferases for enzymatic synthesis of oligosaccharides. In addition, sugar nucleotides, donor substrates of glycosyltransferases, could be produced using genetically engineered bacteria. With the improvements in the supply of glycosyltransferases and sugar nucleotides, it became possible to produce oligosaccharides in large-scale.

Recent researches have revealed that oligosaccharide structures on the cell surface possess important functions in the biochemical recognition processes *(1,2)*. Therefore the applications of oligosaccharides as pharmaceuticals in the fields of the prevention of infections by pathogens, neutralization of toxins, regulations of inflammation, and cancer immunotherapy have been widely recognized *(2)*. However, it has been very difficult to synthesize oligosaccharides, and only very limited amount of oligosaccharides can be obtained by extraction or chemical synthesis.

A lot of chemical and enzymatic methods for synthesis of oligosacchrides have been developed *(3-5)*. Chemical synthesis of oligosaccharides requires multiple protection and de-protection steps, and this complexity does not render the chemical synthesis as realistic as an industrial method. On the other hand, enzymatic synthesis using glycosyltransferases of the Leloir pathway could circumvent the drawbacks of the chemical methods. Glycosyltransferases involved in the biosynthesis of oligosaccharides showed highly stereo- and regiospecific bond formation, and almost no side products were formed during the reactions without protections.

The preparation of glycosyltransferases has been a big problem, however recent progresses in genetic engineering made several glycosyltransferases available in large quantities. The other problem lied in the preparation of sugar nucleotides, substrates of glycosyltransferases, because sugar nucleotides had been very expensive and not readily available. But large-scale synthesis of sugar nucleotides could be possible by the use of bacterial activities.

This article focuses on the recent progress in the production of oligosaccharides using genetically engineered bacteria.

Bacterial Glycosyltransferases

Many mammalian glycosyltransferases have been isolated and most of them were shown to be available for oligosaccharides synthesis. However, most of the genes of mammalian glycosyltransferases were difficult to be expressed in *Escherichia coli*, therefore it restricted the applications of glycosyltransferases for large-scale synthesis of oligosaccharides. Bacterial glycosyltransferase genes are supposed to be attractive sources for glycosyltransferase because of their ready expression in *E. coli.* Due to the continuous researches for microbial genome analysis as well as the progress in screening and cloning techniques, various bacterial genes of glycosyltransferases have been cloned from mainly pathogenic bacteria and most of the genes could be expressed as soluble and active proteins in *E. coli*.

Examples of bacterial glycosyltransferases whose activities were detected in *E. coli* are shown in the Table. The genes of galactosyltransferase were cloned from *Neisseria gonorrhoeae (6)*, *N. meningitidis (7,8)*, *Helicobacter pylori*

(9,10), *Streptococcus pneumoniae (11)*, *S. agalactiae (12)* and *Campylobacter jejuni (13)*. Among them, galactosyltransferases of *N. gonorrhoeae* and *N. meningitidis* were highly expressed in *E. coli*, and utilized for the synthesis of oligosaccharides containing galactose. The genes of *N*-acetylglucosaminyltransferases were also cloned from *N. gonorrhoeae (6)*, *N.*

Table Bacterial Glycosyltransferase

Enzyme	*Source*	*Reference*
GalT		
β1,4 GalT	*Neisseria gonorrhoeae, N. meningitidis*	*(6-8)*
β1,4 GalT	*Helicobacter pylori*	*(9,10)*
β1,4 GalT	*Streptococcus pneumoniae*	*(11)*
β1,4 GalT	*Streptococcus agalactiae*	*(12)*
β1,3 GalT	*Campylobacter jejuni*	*(13)*
β1,3 GalT	*Streptococcus agalactiae*	*(14)*
α1,4 GalT	*Neisseria gonorrhoeae, N. meningitidis*	*(6,7)*
GlcNAcT		
β1,3 GlcNAcT	*Neisseria gonorrhoeae, N. meningitidis*	*(6,7)*
β1,4 GlcNAcT	*Campylobacter jejuni*	*(13)*
GalNAcT		
β1,3 GalNAcT	*Neisseria gonorrhoeae, N. meningitidis*	*(6,7)*
β1,3 GalNAcT	*Campylobacter jejuni*	*(13)*
SiaT		
α2,3 SiaT	*Neisseria gonorrhoeae, N. meningitidis*	*(15)*
α2,3 SiaT	*Campylobacter jejuni*	*(13)*
α2,3 SiaT	*Haemophilus influenzae*	*(16,17)*
α2,3 SiaT	*Haemophilus ducrey*	*(18)*
α2,3 SiaT	*Streptococcus agalactiae*	*(19)*
α2,3/8 SiaT	*Campylobacter jejuni*	*(13)*
α2,6 SiaT	*Photobacterium damsela*	*(20)*
α2,8/9 SiaT	*Escherichia coli*	*(21)*
FucT		
α1,3 FucT	*Helicobacter pylori*	*(22,23)*
α1,2 FucT	*Helicobacter pylori*	*(24)*
α1,3/4 FucT	*Helicobacter pylori*	*(25)*

meningitidis (7) and *C. jejuni (13)*. The genes of sialyltransferase were clone from *N. gonorrhoeae (15)*, *N. meningitidis (15)*, *C. jejuni (13)*, *Photobacterium damsela (20)*, and *E. coli (21)*. In addition, sialyltransferase from *C. jejuni* had a unique characteristic, that it could transfer NeuAc to O-3 of Gal and O-8 of NeuAc which linked with Gal in α2, 3-linkage *(13)*. Three sialyltransferases were identified in *Haemophilus influenzae (16,17)*. Fucosyltransferases were cloned from *H. pylori (22-25)*. In the near future, more and various glycosyltransferases will be cloned with the progress of the analysis of bacterial genomes.

Oligosaccharide Synthesis with Cofactor Regeneration

For the preparation of oligosaccharides using glycosyltransferases, efficient enzymatic production systems were developed using *in situ* regeneration of sugar nucleotides with pyruvate kinase in order to escape from the product inhibition of the glycosyltransferases by the resulting nucleoside diphosphates or monophosphate *(26-28)* (Fig. 1).

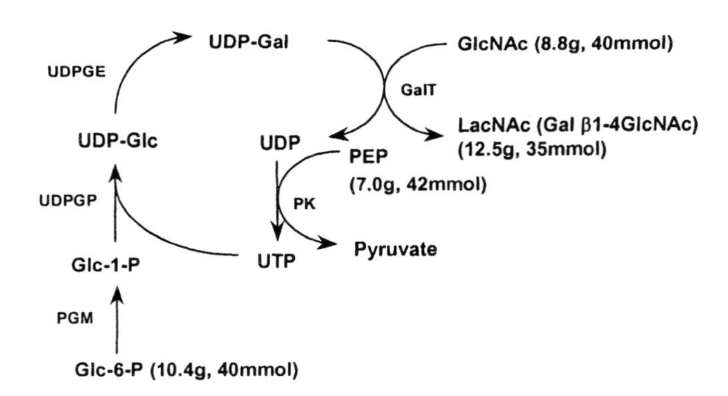

Fig. 1 Synthesis of N*-acetyllactosamine with UDP-Gal recycling (26).*

GalT: β1,4-galactosyltransferase; PGM: phosphoglucomutase; PK: pyruvate kinase; UDPGE: UDP-Gal 4'-epimerase; UDPGP: UDP-Glc pyrophosphorylase: PEP: phosphoenolpyruvate

For the synthesis of sialylated oligosaccharides, a fusion enzyme consisting of CMP-NeuAc synthetase and α2,3-sialyltransferase was created *(29)*. The fusion enzyme was more stable than the original one and could be easily purified. 3'-sialyllactose which exists in human milk was synthesized at a scale of 100 g using the fusion enzyme *(29)*.

Other than the recycling system using pyruvate kinase and phosphoenolpyruvate (PEP) (Fig. 1), sugar nucleotides recycling systems using sucrose synthetase *(30,31)* (Fig. 2) or an inexpensive kinase system with polyphosphate kinase and polyphosphate were also developed *(32)*.

Using bacterial glycosyltransferases which can be easily obtained through the expression in *E. coli*, large-scale synthesis of oligosaccharides with cofactor recycling can be carried out. However it might have limitation because these methods require purified enzymes such as pyruvate kinase and expensive substrates such as PEP.

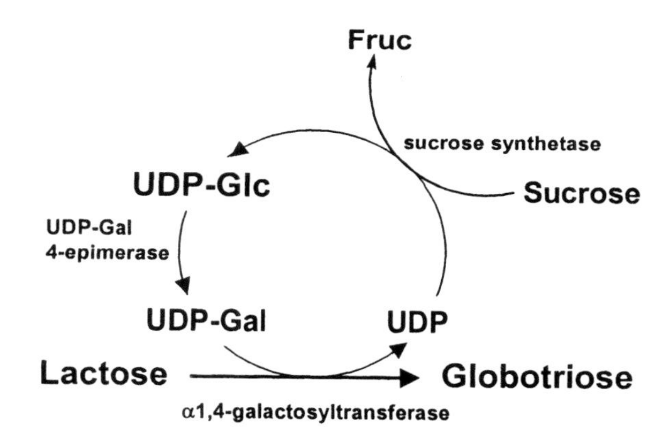

Fig. 2 Oligosaccharides synthesis with sugar nucleotide recycling by sucrose synthetase.

Production of Sugar Nucleotides by Bacterial Coupling

The method of synthesis ATP from adenine using *Corynebacterium ammoniagenes* cells as enzyme sources, which was free from enzyme purification, has been developed *(33)*. In this system, cells were permeabilized by the treatment with surfactant and organic solvent. Phosphoribosylpyrophosphate and ATP were supplied by the metabolism of the cells from glucose. In the same manner, CDP-choline was produced by the combination of *C. ammoniagenes* and recombinant *E. coli* expressing the genes involved in the biosynthesis of CDP-choline *(34)*.

These microbial methods were applied to the synthesis of sugar nucleotides. For instance, UDP-Gal was efficiently produced by the combination of recombinant *E. coli* and *C. ammoniagenes (35)*. Recombinant *E. coli* overexpressed the UDP-Gal biosynthetic genes *galT*, *galK*, *galU*, and *C. ammoniagenes* contributed to the formation of UTP from orotic acid, an inexpensive precursor of UTP. UDP-Gal were accumulated to 44 g/L after 21 h reaction starting with orotic acid and galactose *(35)* (Fig. 3).

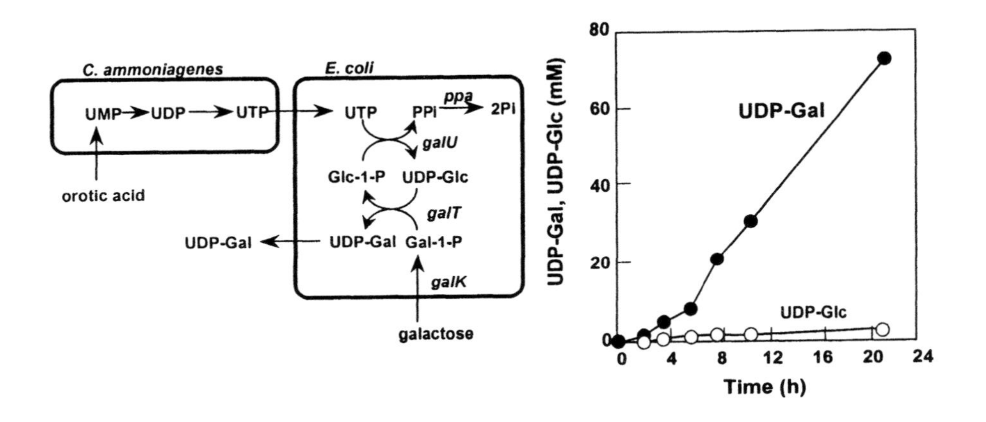

Fig. 3 Production of UDP-Gal by bacterial coupling (35).

galK: galactokinase; galT: Gal-1-P uridyltransferase; galU: UDP-Glc pyrophosphorylase; ppa: pyrophosphatase
(Reproduced with permission from reference 35. Copyright 1998.)

Similarly, 17 g/L of CMP-NeuAc was produced after 27 h reaction starting with orotic acid and NeuAc through the coupling of recombinant *E. coli* cells overexpressing the genes of CMP-NeuAc synthetase and CTP synthetase, and *C. ammoniagenes (36)*.

In the case of GDP-Fuc production, recombinant *E. coli* expressing the genes involved in the biosynthesis of GDP-Fuc was used coupling with *C. ammmoniagenes*. In order to overcome the inhibition of GDP-Fuc on GDP-Man dehydratase, two-step reaction which consisted of the formation of GDP-4-keto-6-deoxymannnose (GKDM) and the conversion of GKDM to GDP-Fuc, was proposed. GDP-Fuc was accumulated at a level of 18.4 g/L after 22 h from GMP and mannose *(37)* (Fig. 4).

By the combination of *C. ammmoniagenes* and *E. coli* expressing the genes involved in sugar nucleotide biosynthesis, a large amount of sugar nucleotides could be accumulated in the reaction mixture.

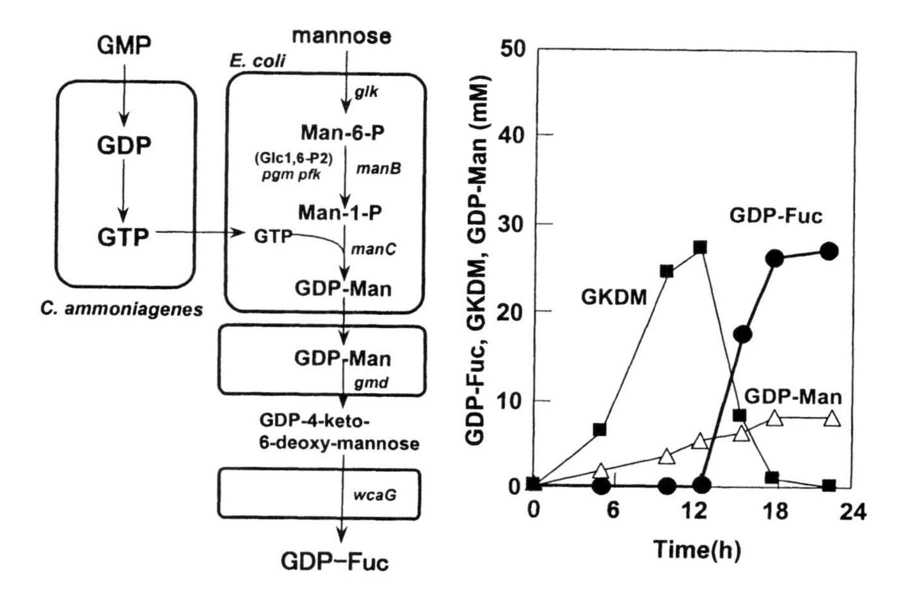

Fig. 4 Production of GDP-Fuc by bacterial coupling (37).

glk: glucokinase; manB: phosphomannomutase; manC: mannose-1-P guanyltransferase; gmd: GDP-mannose dehydratase; wcaG: GKDM epimerase/reductase; pgm: phosphoglucomutase; pfk: phosphofructokinase (Reproduced with permission from reference 37. Copyright 2000 Springer-Verlag.)

Production of Oligosaccharides by Bacterial Coupling

A system of oligosaccharide production by the combination of a bacterial glycosyltransferase and sugar nucleotide production was proposed on the basis that sugar nucleotides could be synthesized by bacterial coupling.

By coupling the recombinant *E. coli* overexpressing the α1,4-galactosyltransferase gene from *N. gonorrhoeae* and the UDP-Gal production system, globotriose (Galα1-4Galβ1-4Glc), which is oligosaccharide portion of the receptor of vero toxin produced by *E. coli* O157, was accumulated to 188 g/L after 36 h from orotic acid, galactose and lactose *(35)* (Fig. 5). When *E. coli* cells overexpressing β1,4-galactosyltransferase gene of *H. pylori* were introduced into the above system, *N*-acetyllactosamine (Galβ1-4GlcNAc) was accumulated to 60 g/L after 20 h *(38)*.

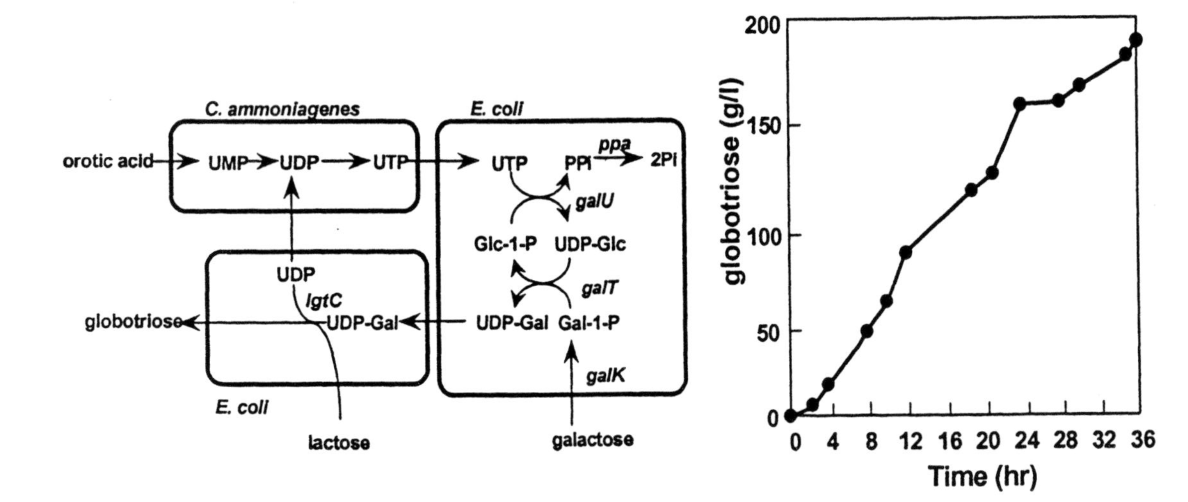

Fig. 5 Production of globotriose by bacterial coupling (35).

galK: galactokinase; galT: Gal-1-P uridyltransferase; galU: UDP-Glc pyrophosphorylase; ppa: pyrophosphatase; lgtC: α1,4-galactosyltransferase (Reproduced with permission from reference 35. Copyright 1998.)

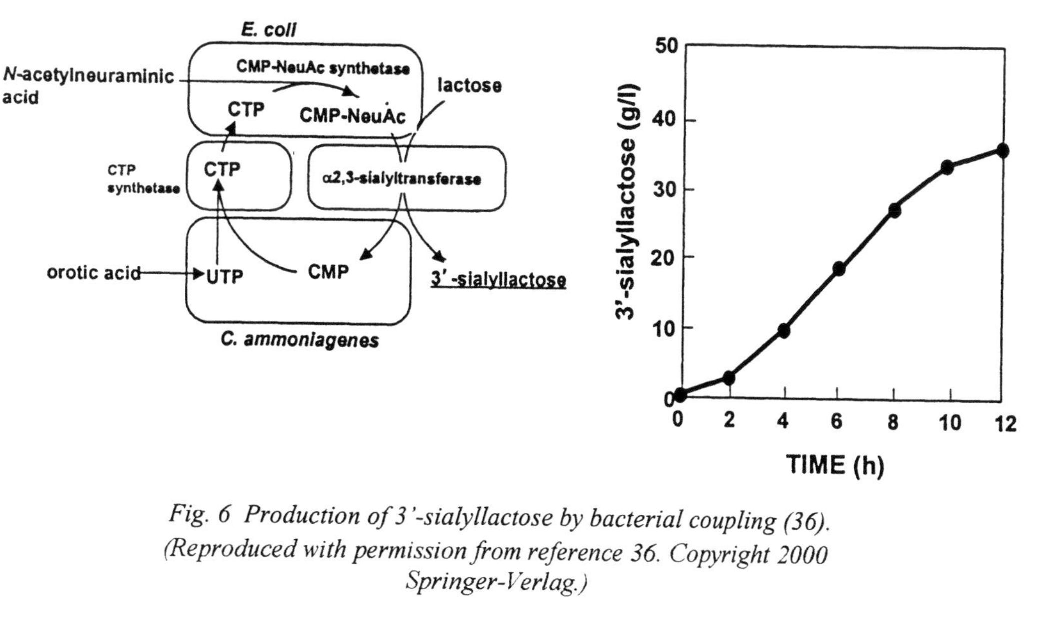

Fig. 6 Production of 3'-sialyllactose by bacterial coupling (36). (Reproduced with permission from reference 36. Copyright 2000 Springer-Verlag.)

Similarly, by the combination of CMP-NeuAc production system and *E. coli* overexpressing the α2,3- sialyltransferase of *N. gonorrhoeae*, 33 g/L of 3'-sialyllactose was produced after 11 h starting with orotic acid, NeuAc, and lactose *(36)* (Fig. 6). In the same way, 45 g/L of sialylTn oligosaccharide (NeuAcα2-6GalNAc), one of cancer associated antigens, was accumulated in 25 h using the α2,6- sialyltransferase of *P. damsela (39)*.

When *E. coli* cells overexpressing α1,3-fucosyltransferase gene of *H. pylori* were introduced to the GDP-Fuc production system, Lewis X (Galβ1-4(Fucα1-3)GlcNAc) was accumulated to 21 g/L after 30 h *(37)*.

By coupling of the sugar nucleotides production system with bacterial glycosyltransferase, an oligosaccharide production system that could be applied to the industrial manufacture was successfully constructed. Oligosaccharides production by bacterial coupling is suitable for large-scale synthesis.

Future Perspective

Large-scale production of oligosaccharides became possible by the use of bacterial glycosyltransferases and the improvement of the supply of sugar nucleotides. Synthesis of oligosaccharides by bacterial coupling is attractive method with high yield, time- and cost- effectiveness, and simple operations. It is expected that biological functions of oligosaccharides will be clarified by sufficient supplies of carbohydrates at a low cost.

References

1. Gagneux, P. and Varki, A. *Glycobiology* **1999,** *9*, 747-755.
2. McAuliffe, J.C. and Hindsgaul, O. *Molecular and Cellular Glycobiology* **2000**, 249-285.
3. Flowers, H.M. *Methods Enzymol* **1978,** *50*, 93-121.
4. Palcic, M.M. *Curr. Opin. Biotechnol* **1999,** *10*, 616-624.
5. Vic, G. and Crout, D.H.G. *Curr. Opin. Chem. Biol.* **1998,** *2*, 98-111.
6. Gotschlich, E.C. *J. Exp. Med.* **1994,** *180*, 2181-2190.
7. Jennings, M.P., Hood, D.W., Peak, I.R.A., Virji, M., and Moxon, E.R. *Mol. Microbiol.* **1995,** *18*, 729-740.
8. Wakarchuk, W.W., Martin, A., Jennings, M.P., Moxon, E.R., and Richards, J.C. *J. Biol. Chem.* **1996,** *271*, 19166-19173.

9. Logan, S.M., Conlan, J.W., Monteiro, M.A., Wakarchuk W.W., and Altman, E. *Mol. Microbiol.* **2000,** *35*, 1156-1167.
10. Endo, T., Koizumi, S., Tabata, K., and Ozaki, A. *Glycobiology* **2000,** *10*, 809-813.
11. Kolkman, M.A.B., Wakarchuk, W.W., Nuijten, P.J.M., and van der Zeijst, B.A.M. *Mol. Microbiol.* **1997,** *26*, 197-208.
12. Yamamoto, S., Miyake, K., Koike, Y. Watanabe, M., Machida, Y., Ohta, M., and Iijima, S. *J. Bacteriol.* **2000,** *181*, 5176-5184.
13. Gilbert, M., Brisson, J.R., Karwaski, M.F., Michniewicz, J., Cunningham, A.M., Wu, Y., Young, M., and Wakarchuk, W.W. *J. Biol. Chem.* **2000,** *275*, 3896-3906.
14. Watanabe, M., Miyake, K., Yanae K., Kataoka, Y., Koizumi, S., Endo, T., Ozaki, A., and Iijima, S. *J. Biochem.* **2002,** *131*, 183-191.
15. Gilbert, M., Watson, D.C., Cunningham, A.M., Jennings, M.P., Young, N.M., and Wakarchuk, W.W. *J. Biol. Chem.* **1996,** *271*, 28271-28276.
16. Hood, D.W., Cox, A.D., Gilbert, M., Makepeace, K., Walsh, S., Deadman, M.E., Cody, A., Martin, A., Mansson, M., Schweda, E.K.H., Brisson, J.-R., Richards, J.C., Moxon, E.R., and Wakarchuk, W.W. *Mol. Microbiol.* **2001,** *39*, 341-350.
17. Jones, P.A., Samuels, N.M., Phillips, N.J., Munson, R.S. Jr., Bozue, J.A., Arseneau, J.A., Nichols, W.A., Zaleski, A., Gibson, B. W., and Apicella, M.A. *J. Biol. Chem.* **2002,** *277*, 14598-14611.
18. Bozue, J.A., Tullius, M.V., Wang, J., Gibson, B.W., and Munson, R.S. Jr. *J. Biol. Chem.* **1999,** *274*, 4106-4114.
19. Chaffin, D.O., McKinnon, K., and Rubens, C.E. *Mol. Microbiol.* **1996,** *45*, 109-122.
20. Yamamoto, T., Nakashizuka, M., and Terada, I. *J. Biochem.* **1998,** *123*, 94-100.
21. Shen, G.J., Datta, A.K., Izumi, M., Koeller, K.M., and Wong, C.-H. *J. Biol. Chem.* **1999,** *274*, 35139-35146.
22. Martin, S.L., Edbrook, M.R., Hodgman, T.C., van den Eijnden, D.H., and Bird, M.I. *J. Biol. Chem.* **1997,** *272*, 21349-21356.
23. Ge, Z., Chan, N.W.C., Palcic, M.M, and Taylor, D.E. *J. Biol. Chem.* **1997,** *272*, 21357-21363.
24. Wang, G., Boulton, P.G., Chan, N.W.C., Palcic, M.M, and Taylor, D.E. *Microbiol.* **1999,** *145*, 3245-3253.
25. Rasko, D.A., Wang, G, Palcic, M.M, and Taylor, D.E. *J. Biol. Chem.* **2000,** *275*, 4988-4994.
26. Wong, C.-H., Haynie, S.L., and Whitesides, G.M. *J. Org. Chem.* **1982,** *47*, 5416-5418.
27. Ichikawa, Y., Shen, G.-J., amd Wong, C.-H. *J. Am. Chem. Soc.* **1991,** *113*, 4698-4700.

28. Ichikawa, Y., Look, G.C., and Wong, C.-H. *Anal. Biochem.* **1992,** *202*, 215-238.
29. Gilbert, M., Cunningham, A.M., DeFrees, S., Gao, Y., Watson, D.C., Young, N.M., and Wakarchuk, W.W. *Nat. Biotechnol.* **1998,** *16*, 769-772.
30. Hokke, C.H., Zervosen, A., Elling, L., Joziasse, D.H., and van den Eijnden. *Glycoconj. J.* **1996,** *13*, 687-692.
31. Chen, X., Zhang, J., Kowal, P., Liu, Z., Andeana, P.R., Lu, Y., and Wang, P.G. *J. Am. Chem. Soc.* **2001,** *123*, 8866-8867.
32. Noguchi, T. and Shiba, T. *Biosci. Biotechnol. Biochem.* **1998,** *62*, 1594-1596.
33. Fujio, T. and Furuya, A. *J. Ferment. Technol.* **1983,** *61*, 261-267.
34. Fujio, T. and Maruyama A. *Biosci. Biotechnol. Biochem.* **1997,** *61*, 856-959.
35. Koizumi, S., Endo, T., Tabata, K., and Ozaki, A. *Nat. Biotechnol.* **1998,** *16*, 847-850.
36. Endo, T., Koizumi, S., Tabata, K., and Ozaki, A. *Appl. Microbiol. Biotechnol.* **2000,** *53*, 257-261.
37. Koizumi, S., Endo, T., Tabata, K., Nagano, H., Ohnishi, J., and Ozaki, A. *J. Ind. Micriobiol. Biotechnol.* **2000,** *25*, 213-217.
38. Endo, T., Koizumi, S,. Tabata, K., and Ozaki, A. *Carbohydr. Res.* **1999,** *316*, 179-183.
39. Endo, T., Koizumi, S,. Tabata, K., and Ozaki, A. *Carbohydr. Res.* **2001,** *330*, 439-443.

Chapter 12

Synthesis of Glycoconjugates through Biosynthesis Pathway Engineering

Mei Li[1], Jun Shao[1], Min Xiao[2], and Peng George Wang[1,3]

[1]Department of Chemistry, Wayne State University, Detroit, MI 48202
[2]Department of Microbiology, School of Life Science, Shandong University, Jinan, Shandong 250100, Peoples Republic of China
[3]Current address: Departments of Chemistry and Biochemistry, The Ohio State University, 876 Biological Sciences Building, 484 West 12th Avenue, Columbus, OH 43210

The synthesis of biologically important glycoconjugates through biosynthesis pathway engineering is more intensively investigated in the past few years. Many stretagies have been demonstrated suitable for large scale synthesis of oligosaccharides using either glycosyltransferases isolated from engineered bacteria or whole cells as enzyme source. These advances contribute a great in glycobiology science.

Introduction

Glycobiology has become an intensive and fast growing research area due to the fact that carbohydrates and glycoconjugates including glycolipids and glycoproteins, play essential roles in the cell biological functions such as cellular recognition, signal transduction, tumor metastasis, and immune responses *(1-5)*. A limiting factor in developing carbohydrate-based compounds for clinical application is the high cost and complexity of producing glycans in adequate quantities. Many efforts have been invested in the synthesis of oligosaccharides. They fall into two general strategies: chemical synthesis and enzymatic synthesis. The chemical synthesis of oligosaccharides involves in tedious steps of protection and deprotection to obtain desired stereochemistry and regiochemistry. On the contrary, enzymatic approaches employing isolated enzymes, glycosyltransferases and glycosidases, or engineered whole cells are highly regio-and stereo-selective, and are often highly efficient, which makes enzymatic approaches more attractive, powerful and complimentary to chemical methods alone *(6)*.

Two classes of enzymes are available for the enzymatic synthesis of glycosides: glycosidases and glycosyltransferases. Glycosidases are often readily available and use simple glycosyl donors. However, glycosidases usually catalyze the reverse reaction and are generally not as regioselective as transferases *(7)*. In comparison, glycosyltransferases can perform glycosylation reactions in high yields with excellent selectivity. And, therefore, glycosytransferases are especially useful for the efficient synthesis of complex oligosaccharides and glycoconjugates *(8)*. However, normally the glycosyltransferases are not readily available. In addition, sugar-nucleotide, which serves as an a donor substrate in enzymatic glycosylation, are prohibitively expensive. Advances in cost-effective technology for enzymatic synthesis appear to be promising solutions to this problem. This approach uses recombinant glycosyltransferases, relatively inexpensive precursors, and recycling reactions to regenerate most costly sugar-nucleotides *(9,10)*. Herein, our discussion mainly focuses on using biosynthesis pathway engineering approach for practical synthesis of oligosaccharides with special emphasis on some successful examples in our research group.

Sugar Nucleotide Synthesis

The glycosyltransferase requires a sugar nucleotide as substrate. To prepare oligosaccharides efficiently, in situ regeneration system for common sugar nucleotides have been developed by mimicing the natural biosynthetic pathways. There are nine common sugars seen in natural oligosaccharides and glycoconjugates. Our group has developed an efficient synthesis of eight common sugar nucleotides: UDP-Glc, UDP-Gal, UDP-GlcNAc, UDP-GalNAc, UDP-GlcA, GDP-Man, GDP-Fuc and CMP-NeuAc, and has applied them in synthesis carbohydrates employing specific glycosyltransferases.

UDP-Glc and UDP-Gal

Following the known biosynthetic pathway shown in Scheme 1, UDP-Glc can be regenerated by sucrose synthase and applied into glycosylation. On the other hand, UDP-Gal regeneration system was established in two approaches based on two different biosynthetic pathways. By adding UDP-Glc-4-epimerase (GalE, EC 5.1.3.2) into the sucrose synthase-catalyzed UDP-Glc regeneration cycle, UDP-Gal can be regenerated in an efficient way (Scheme 9a). The key enzyme is the sucrose synthase (*SusA*) cloned from cyanobacterium *Anabaena sp.*PCC 7119. Another approach starts from direct phosphorylation at the 1-position of galactose *(10)*, to give Gal-1-P, which can be converted to UDP-Gal via an uridyltransferase with UDP-Glc (Scheme 7). Both UDP-Gal regeneration systems were applied into "superbug" (metabolically engineered bacteria) and validated in the synthesis of α-Gal trisaccharide and Globotriose *(11,12)*.

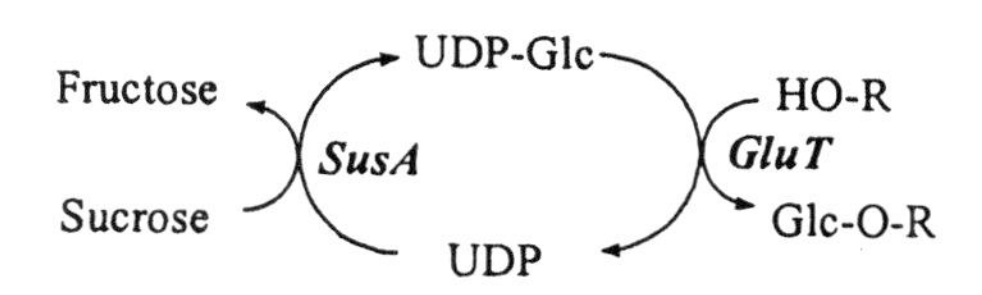

Scheme 1 Regeneration of UDP-Glc

UDP-GlcNAc and UDP-GalNAc

Following the biosynthetic pathway of synthesizing UDP-GlcNAc in eukaryotes (Scheme 2), GlcNAc is first phosphorylated by GlcNAc kinase (GlcNAcK, *C. albicans*) *(13)*. Then GlcNAc phosphatemutase (Agm1, *S. cerevisiae*) *(14)* converts GlcNAc-6-P to GlcNAc-1-P, which is subsequently uridylated to form UDP-GlcNAc by a truncated UDP-GlcNAc pyrophosphorylase (GlmU, *E. coli*) *(15)*. The resulting ADP can be reconverted to ATP by pyruvate kinase (PykF, *E. coli*) *(16)* with the consumption of one equivalent of phosphoenolpyruvate (PEP). The by-product pyrophosphate (PPi) is finally hydrolyzed by inorganic pyrophosphatase (PPA, *E. coli*) *(17)*. An agarose bead co-immobilized with mutiple enzymes (named as "Superbead") was achieved for synthesis UDP-GlcNAc *(18)*.

UDP-GalNAc can be generated by the addition of UDP-GalNAc4-epimerase (GalNAcE, EC 5.1.3.2) to the regeneration cycle for UDP-GlcNAc. We have cloned and overexpressed a novel UDP-GalNAc4-epimerase (wbgU) from *Plesiomonas shigelloide (19)*. Efficient UDP-GalNAc regeneration system (Scheme 3) was established and used in synthesis of Globotetraose and a series of derivatives with applying the recombinant β-1,3-*N*-acetylgalactosaminyltransferase (lgtD) from *Haemophilus influenzae* strain Rd *(20)*.

GlcNAcK *Agm1*
GlcNAc → GlcNAc-6-P → GlcNAc-1-P PP$_i$ —*PPA*→ 2P$_i$
ATP ADP
PykF
Pyruvate PEP
GlmU
UTP UDP-GlcNAc
Pyruvate HO-R
PykF
UDP *LgtA* (β1,3GlcNAcT)
PEP GlcNAc-O-R

GlcNAc + 2PEP + HOR → GlcNAcOR + 2Pyruvate + 2Pi

Scheme 2 UDP-GlcNAc regeneration cycle from GlcNAc

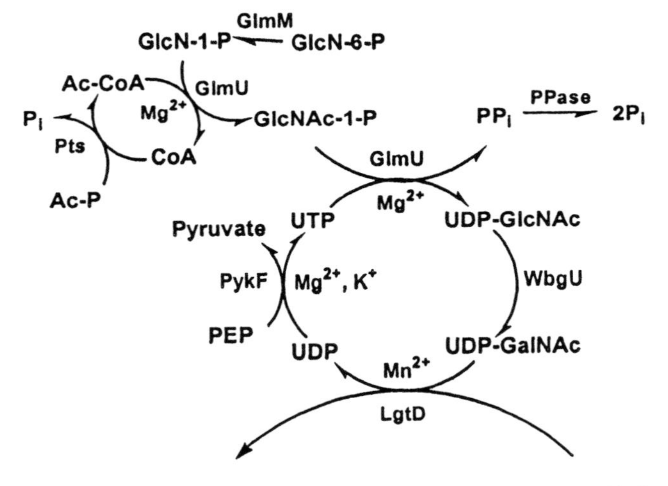

Scheme 3 Synthesis of Globotetraose with UDP-GalNAc regeneration

GDP-Mannose and GDP-Fucose

GDP-Man regeneration system starting from Man-1-P has been report by Wong's group*(9)*. As the Fru-6-P is inexpensive to obtain, we developed a new regeneration pathway from Fru-6-P with all recombinant enzymes. As shown in Scheme 4, the cycle comprises of three key enzymes: D-mannose-6-phosphate isomerase (PMI, EC 5.3.1.8), phosphomannomutase (PMM, EC 5.4.2.8) and

GDP-mannose pyrophosphorylase (GMP, EC 2.7.7.13). The recombinant bifunctional PMI/GMP (phosphomannose isomerase/GDP-D-mannose pyrophosphorylase) from *Helicobacter pylori* has been cloned and overexpressed in *E. coli (21)*. The bifunctional enzyme PMI/GMP catalyzes both the first and third steps of GDP-$_D$-mannose biosynthesis from D-fructose-6-phosphate. In addition, PMM of *H. pylori* was also cloned and overexpressed *(21)*. The system was coupled with a truncated α1,2-mannosyltransferase (ManT catalytic domain, *Saccharomyces cerevisiae)* to synthesize the mannosyloligosaccharides (not published).

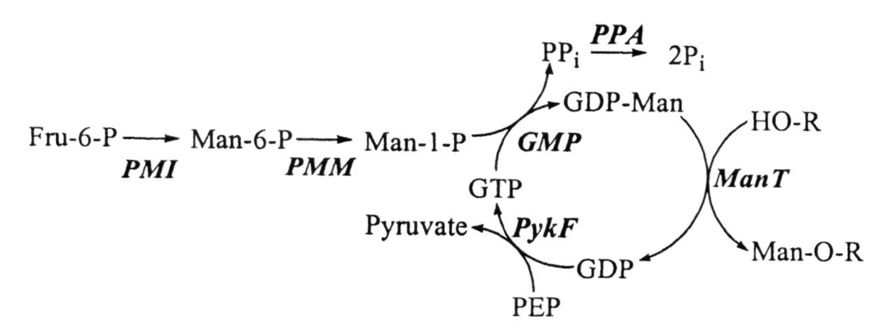

Scheme 4 GDP-Man regeneration cycle

GDP-Fuc regeneration cycle was constructed by incorporating two more enzymes GDP-mannose-4,6-dehydratase (GMD) and GDP-fucose synthase (GFS) into the same biosynthetic pathway of GDP-Man from Fru-6-P (Schem 5). Both GMD and GFS were cloned from *Helicobacter pylori*. With the bifuctional gene PMI/GMP adjacent to them, GMD, GFS and PMI/GMP array in sequence to form the GDP-fucose biosynthesis gene cluster in the chromosome genome of *Helicobacter pylori (22)*.

CMP-Neu5Ac

Based on the existing CMP-Neu5Ac regeneration system reported by Wong's group *(23,24)*, we extended the regeneration using ManNAc as starting material and creatine phosphate as energy source. The CMP-NeuNAc regeneration was constructed with five enzymes as components: NeuAc aldolase (NanA), CMP-Neu5Ac synthetase (NeuA, EC 2.7.7.43), CMP kinase (CMK, Ec 2.7.4.14) and creatine kinase (CK, EC 2.7.3.2) coupled with a new α2,3-

sialyltransferase cloned from *Pasteurella multocida* PM70. As shown in Scheme 6, here creatine phosphate was used as energy source instead of PEP, because creatine phosphate is a more efficient, cheap and convenient energy

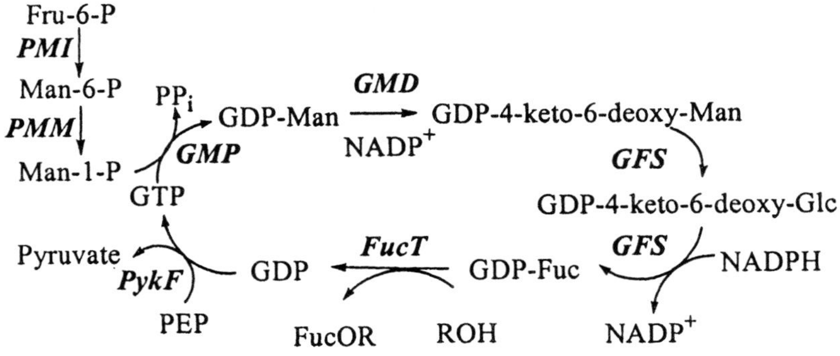

Fru-6-P + PEP + HOR ⟶ FucOR + Pyruvate + PPi

Scheme 5 GDP-Fuc regeneration cycle

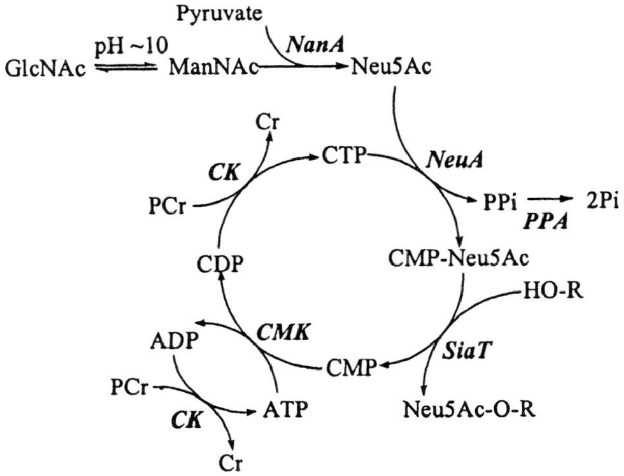

GlcNAc + Pyruvate + 2PCr + HOR ⟶ Neu5AcOR + 2Cr + 2Pi

Scheme 6 CMP-Neu5Ac regeneration cycle

source compared with PEP, especially in large-scale synthesis of gylcoconjugates *(25)*. ManNAc can be epimerized chemically from GlcNAc under basic condition (pH10) *(26)*, thus GlcNAc can also be used as starting material in sialyloligosaccharide synthesis.

Cell Free Oligosaccharide Synthesis

Glycosyltransferases catalyze the transfer of a monosaccharide from a donor, corresponding sugar nucleotide, to saccharide acceptors. With the aid of genetic technology and the availability of entire genome sequences, numerous enzymes are being isolated and overexpressed heterologously and subsequently used in the synthesis of oligosacchrides. In the early 90's, Wong's group established one-pot system, the first one brought up multiple-enzyme sugar nucleotide regeneration system for carbohydrate synthesis. This system was successfully used in the synthesis of lactose, N-acetyllactosamine (LacNAc), sialylLacNAc and their derivatives with in situ cofactor regeneration, in which the sugar nucleotides are continuously regenerated from readily available monosaccharides*(10,23,24)*. The multienzyme systems were demonstrated to be operated very efficiently without problems of product inhibition, for example, the sialy-LacNAc was obtained in 89% yield using LacNAc as acceptor with in situ CMP-sialic acid regeneration *(23)*. However, in this one-pot system, purified enzymes were used and the recombinant enzyme remained too expensive. Moreover the enzyme could not be reused and it is not convenient to separate the product from the system. These factors are not suitable for large-scale synthesis of oligosaccharides. Alternatively, immobilized enzyme system has advantages such as ease of product separation, increased stability and reusability of the catalysts. Thereby, much more research efforts focused on immobilization of glycosyltransferases and other enzymes coupling with the water soluble polymer supports to circumvent these limitations*(6,27-32)*.

Wong's and Nishimura's method

Recently two applications of immobilized multienzyme coupling with aqueous polymer supports in synthesis of oligosaccharides have been reported *(27,28)*. Nishimura et al. made use of glycosyltransferases as fusion proteins with maltose-binding protein (MBP). The function of MBP domain is act as a specific affinity tag for both purification and immobilization of this engineered biocatalyst as well as solubilization. In addition, the acceptor was immobilized on water-soluble polymer involving an α-chymotrypsin-sensitive linker moiety. They examined the feasibility of the immobilized MBP-β1,3-GlcNAcT adsorbed on amylase resin prepared on the basis of the specific sugar-protein interaction without any chemical treatment of cross-linking. About 80% of transfer of

GlcNAc residue from UDP-GlcNAc toward LacCer polymer was achieved *(28)*. On the other hand, Wong's group developed homogenous enzymatic synthesis system using thermo-responsive water-soluble polymer support. Several enzymes immobilized on thermo-responsive polyacrylamide polymers are nearly as active as their soluble forms and can be recovered for reuse after gentle heating and precipitation. The trisaccharide Le^X was synthesized in 60% yield with no chromatographic purification of intermediates *(27)*.

Wang's Superbeads

Wang's group reported a new approach that transfers in vitro multiple enzyme sugar nucleotide regeneration systems onto solid beads (the Superbeads) which can be reused as common synthetic reagents for production of glycoconjugates *(33)*. The beads were prepared in following steps: (i) cloning and overexpression of individual N-terminal His_6-tagged enzymes along the sugar nucleotide biosynthetic pathway and (ii) co-immobilizing these enzymes onto nickelnitrilotriacetate (NTA) beads. The sugar nucleotide regeneration superbeads can be conveniently combined with glycosyltransferases either on beads or in solution for specific oligosaccharide sequences. We developed first generation of superbeads for UDP-Gal regeneration along with biosynthesis pathway (Scheme 7a). As shown in Scheme 7, the required enzymes for UDP-Gal regeneration: galactokinase (GalK), galactose-1-phosphate uridylyltransferase (GalPUT), glucose-1-phosphate uridylyltransferase (GalU), and pyruvate kinase (PykF), were immobilized quantitively onto the agarose beads (Ni^{2+}-NTA resins) because all the enzymes were recombinant enzymes with His_6-tag. The feasibility was demonstrated by gram scale synthesis of Galα1,3Gal,1,4GlcOBn with a truncated bovine α-1,3-galactosyltransferase (α1,3GalT) expressed in *E. coli*. A yield of 72% was achieved based on acceptor LacOBn. After three times reuse (yields were 71%, 69%, and 66%, respectively) within three weeks, 90% enzyme activity of α1,3GalT retained. The versatility of the UDP-Gal regeneration beads was demonstrated in the synthesis of a variety of oligosaccharides such as LacNAc (Galβ1,4GlcNAc, 92% yield) and globotriose Gb3 (Galα1,4Galβ1,4Glc, 86% yield). In addition, we also prepared superbeads used to produce UDP-GlcNAc. A gram-scale synthesis of UDP-GlcNAc was conducted at 30°C and a 50% yield of product can still be achieved after five 20 h reaction cycles *(18)*. After repeated reactions, the deactivated enzymes could be removed from the nickel agarose beads and be recharged for further uses.

In summary, the superbeads possess advantages of ease of separation, increased stability and reusability. The superbeads may become a new generation of bio-reagents that can be coupled with a variety of glycosyltransferases for the production of glycoconjugates and their derivatives.

(a)

PEP, ATP, Galactose, K^+, Mg^{2+}, PykF, GalK, Pyruvate, ADP, Gal-1-P, GalT, UDP-Gal, Galβ1,4Glc, UDP-Glc, Mg^{2+}, Glc-1-P, Mn^{2+}, α1,3GalT, PPi, GalU, UTP, K^+, Mg^{2+}, UDP, Galα1,3Lac, PykF, Pyruvate, PEP

$$\text{Gal} + \text{Lac} + 2\text{PEP} \longrightarrow \text{Gal}\,\alpha 1{,}3\text{Lac} + 2\text{Pyruvate} + \text{PPi}$$

GalK: Galactokinase, GalT: Gal-1-P uridylyltransferase, PykF: Pyruvate kinase, GalU: Glc-1-P uridylyltransferase, α1,3GalT: α1,3-galactosyltransferase

(b)

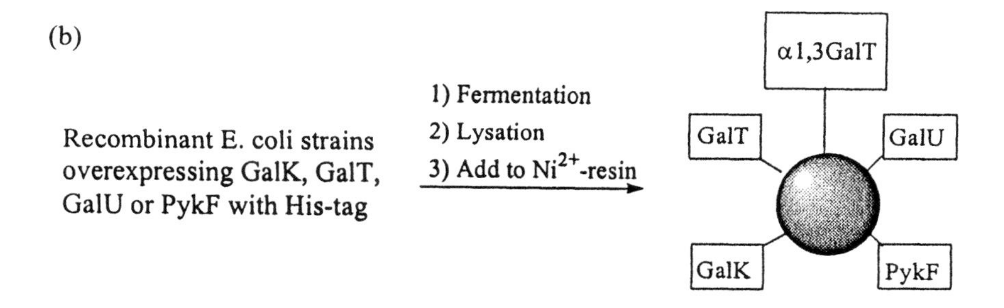

Scheme 7 Superbeads and biosynthetic pathway of α-Gal trisaccharide with regeneration of UDP-Gal

Genetically Engineered Bacteria : Whole Cell Approaches

As the isolation of enzymes from engineered cells is, in general, a laborious operation and the purification process may result in decreased enzymatic activity. A rapidly emerging method for large-scale biocatalytic production of oligosaccharides is the use of microorganism metabolically engineered. There is no need to isolate enzymes and biotransformations can be carried out with inexpensive precursors *(6)*. Recently, many approaches have been conducted to produce oligosacchrides through whole-cell process. Kyowa Hakko Inc. in Japan developed a system for large-scale synthesis of oligosaccharides (Scheme 8) by coupling multiple metabolically engineered bacteria *E. coli* with *Corynebacterium ammoniagenes (34-37)*. Kyowa Hakko's system has been successfully applied in large-scale synthesis of globotiose (Galα1,4Galβ1,4Glc, 188g/L) *(34,35)*, 3'-sialyllactose (NeuAcα2,3Galβ1,4Glc, 33 g/L) *(36)*, and Lewis X [Galβ1,4(Fucα1,3)GlcNAc, 21 g/L] *(38)*. However, this system still have some drawbacks, 1) multiple fermentations of several bacterial strains are required, including one engineered *C. ammoniagenes* and two or more engineered *E. coli* strains; 2) the permeabilization of the bacteria with xylene to allow a passive circulation of the substrates between different bacterial strains; 3) the reaction is carried out by nongrowing cells. On the contrary, Wang's group has developed the "superburg" technology utilizing single recombinant bacteria carrying only one engineered recombinant plasmid to perform large-scale synthesis oligosaccharides to overcome the above issues.

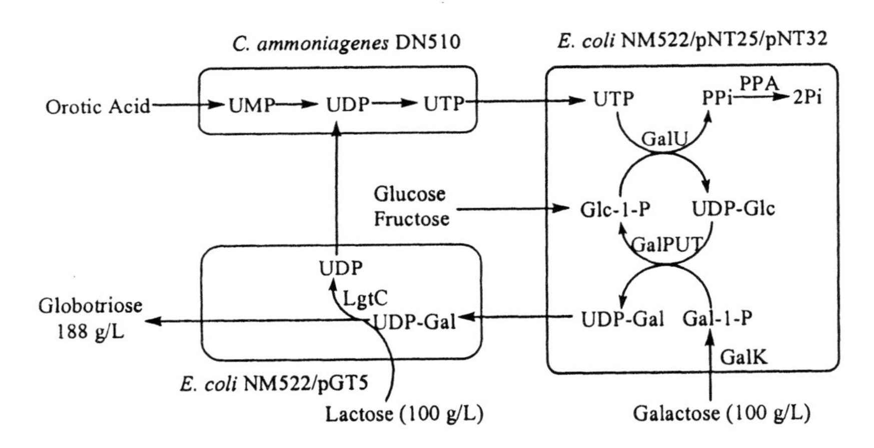

Scheme 8 Kyowa Hakko's technology through multiple bacterial coupling

Wang's superburg

Based on our previous successful research in synthesis carbohydrates utilizing purified recombinant enzymes and superbeads *(18,33,39-41)*, we further explored new approach to synthesize carbohydrates using metabolically engineered whole cells. Different from Kyowa Hakko's approach, we use single engineered bacteria containing one recombinant vector, on which the genes encoding the enzymes along the biosynthetic pathway for sugar-nucleotide regeneration and oligosaccharide accumulation, were assembled to form an artificial gene cluster. Therefore, the engineered bacteria named as "superbug", is capable of synthesis carbohydrate with *in situ* sugar nucleotide regeneration. In this approach, there is no need to purify and immobilize individual enzymes.

The choice of pLDR20 vector enabled convenient temperature induction and eliminated the need for inducer, IPTG.

Our whole-cell catalyzed production of oligosaccharides is a two-step process. The first step involves the growth of the recombinant *E. coli* NM522 cells and the overexpression of the recombinant enzymes along the biosynthetic pathway. In the second step the harvested cells are permeabilized by repeated freeze/thaw and employed as catalysts in the reaction. The simple permeabilization allows the transport of substrates and products into/out of the cells. The permeabilized cells, although not healthy and much less viable, can still carry out certain metabolism (e.g., glycolysis) to provide the necessary bioenergetics to drive the enzymatic reactions inside the cells. Importantly, this two-step process allows the use of high cell concentration in the second step to achieve a high-yield fed-batch process.

Synthesis of α-gal epitope

The "superbug" strategy was validated by the synthesis of biomedically important α-Gal trisaccharides epitope (also called α-Gal epitope). Galα1,3Gal terminated oligosaccharide sequences exist on cell-surface glycolipids or glycoproteins in mammals other than humans, apes, and old world monkeys. It is the major antigen responsible for the hyperacute rejection in pig-to-human xenotransplantation.

In attempt to synthesis α-Gal trisaccharides epitope, a biosynthetic pathway to Galα1,3Lac with recycling of UDP-Gal were designed with five enzymes: GalK, GalT, GalU, PykF, and α1,3-galactosyltransferase (α1,3GT) (Scheme 7a). Each enzyme was overexpressed individually on the pET15b vector to demonstrate the enzyme activities, followed by cloning onto another expression plasmid pLDR20 in sequence to construct the final plasmid pLDR20-αKTUF containing an artificial biosynthetic gene cluster (Fig 1). Since lactose is used as acceptor in the synthesis reaction, the *E. coli* strain NM522, which is a β-galactosidase defective, was used as a host strain for co-expression of the enzymes. Without extensive optimization, the whole cell superbug system produces 3~4 g of Galα1,3Lac trisaccharide (α-Gal trisaccharides) in every 10 L fermentation, starting from inexpensive precursor and only catalytic amounts of ATP and PEP *(11)*.

Synthesis of Globotriose and derivatives

The superbug technology can be expanded when one or more glycosyltransferases are incorporated in the recombinant *E. coli*. Replacing α-1,3GalT with other transferases such as LgtC (α-1,4galactosyltransferase), superbug CKYUF was obtained for synthesis of globotriose and its derivatives.

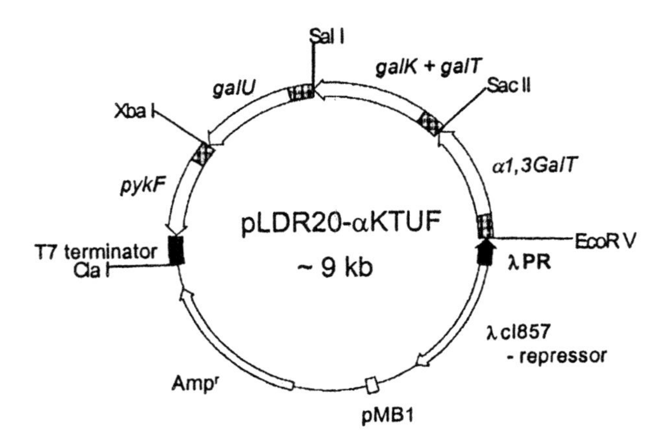

Fig 1 Plasmid map of pLDR20-αKTUF

Alternative superbug for synthesis of galactosides was constructed making use of the simplified cycle of UDP-Gal regeneration based on a sucrose synthase (SusA, EC 2.4.1.13) (Scheme 9) *(42,43)*. The synthesis and cleavage of sucrose is catalyzed by sucrose synthase (sucrose + UDP ⇔ UDP-Glc + fructose). On the basis of this in vitro biosynthetic cycle, the three genes comprising a galactosyltransferase gene (LgtC, α1,4galactosyltransferase) from *Neisseria meningitidis*, *gal*E gene (UDP-galactose 4-epimerase) from *E. coli* K-12, and *sus*A gene from *Anabaena* sp. PCC 7119 (ATCC29151), were assembled into an artificial gene cluster in plasmid pLDR20 resulted in pLDR20-CES. The reactions reached saturation after 36 h. Galα1,4Lac (Gb3) accumulated to 44 mM (22 g/L) *(12)*.

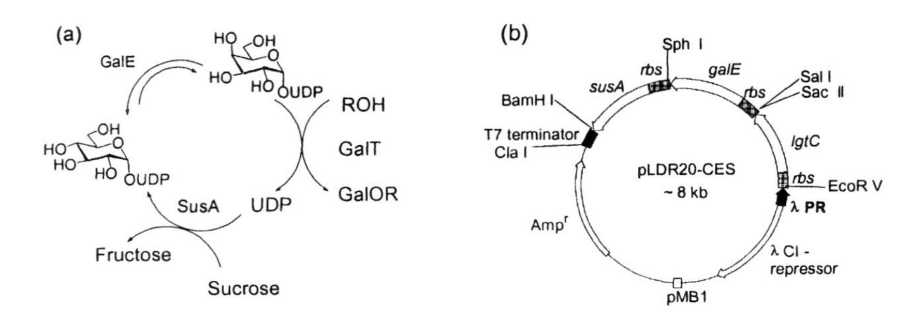

Scheme 9 (a) Biosynthetic Pathway for Galactosides Using Sucrose Synthase (b) Map of pLDR20-CES

The whole-cell system eliminates the need to purify and preserve the enzymes involved. Moreover, the use of a single bacterial strain instead of multiple strains avoids transport of reaction intermediates between strains and the complication of maintaining the growth of different strains. Since the process uses only inexpensive mono- and disaccharides and catalytic amount of UDP (1%), the production can be economically carried out on industrial scale.

Living cell synthesis oligosaccharide

Another attractive strategy for synthesis of oligosaccharides is in vivo, intracellular production in recombinant bacteria expressing glycosyltransferases *(44-46)*. This method possesses some advantages over nonliving cell systems. There is no need for the purification of glycosyltransferases and cells already possess the machinery required for sugar nucleotide synthesis. Samain *et al.* have published several reports on using growing bacteria cells as natural mini-reactors for regeneration of sugar nucleotides and to utilize the intracellular pool of sugar nucleotide as substrate for in vivo synthesis of oligosaccharides. For example, N-acetyl-chitopentaose (2.5 g/L) was produced by cultivating *E. coli* expressing *nodC* genes from *Azorhizobium caulinodans* that encode chitooligosaccharide synthase *(44)*. When *E. coli* cells coexpressing *nodC* from *A. caulinodans* and *lgtB* from *N. meningitidis* were cultivated, more than 1.0 g/L of hexasaccharide, identified as βGal(1,4)[βGlcNAc(1,4)]$_4$GlcNAc, accumulated *(46)*. Recently, they extended the "living factory" system in human milk oligosaccharides synthesis by cultivating the metabolically engineered bacteria. High-cell-density cultivation of lacZ$^-$ strains that overexpressed the β1,3-N-acetylglucosaminyltransferase *lgtA* gene of *Neisseria meningitidis* resulted in the synthesis of 6 g/L of trisaccharide (GlcNAcβ1-3Galβ1-4Glc). When the β1,4-galatocyltransferase *lgtB* gene of *N. meningitidis* was coexpressed with *lgtA*, lacto-N-neotetraose and lacto-N-neoheaxose were produced with a yield higher than 5 g/L *(47)*. In these cases, the oligosaccharides were produced inside the cells. The engineered bacteria were cultivated at high cell density with glycerol as the carbon and energy source using classical fed-batch fermentation without IPTG.

Further Improvement in Metabolically Engineered Bacteria System

The whole cell approach to production of oligosaccharide has been proved to be an attractive method with high yields, time and cost effectiveness, reproducibility and simple operation. However, to reach to the industrial scale synthesis of oligosaccharides economically there are still some aspects need further improvements:

- New expression medium. Currently most molecular biologists are using complex (rich) media such as Luria-Bertani for protein expression. However some problems could evolve from using rich media. The bacteria grow at such a great rate that they tend to be unstable, especially in large-scale production. Therefore it is necessary to develop expression strategy using only minimal medium containing chemically well-defined carbon source, nitrogen source and inorganic salts.
- New selection marker. Up to now the most common selection marker for recombinant bacteria is antibiotic resistance genes such as ampicillin resistance gene et al. But during the fermentation, especially in large scale, antibiotics are quickly cleared from the culture by the enzymes encoded by resistance genes harboring on the multi-copy plasmids. After the depletion of antibiotics from the medium, the growth of bacteria that lost recombinant plasmids will be dominant over the ones that are burdened by the replication of high copy plasmid and the expression of the recombinant genes. In addition, the use of high concentration of antibiotics in large-scale production is not practical.
- New expression strategy. The target genes will either be integrated into chromosome or carried in a plasmid. The genes integrated into the chromosome will be those responsible for the synthesis of sugar nucleotide and those for energy supply. The plasmid will also contain new selectable marker.

Development of new generation superbug for synthesis sialyllactose is undergoing in Wang's group to overcome the above problems (Fig. 2). Serine auxotrophy is used as selection marker since the superbug will be cultivated in minimal medium. Site directed integration of recombinant genes into *serA* gene on the *E.coli* chromosome resulted in disruption of *serA* (3-phosphoglycerate dehydrogenase) via suicide plasmid pMAK705. The growth of "superbug" will depend on the maintenance of a *serA* locus carried on a plasmid. In summary, the integrated genes will provide stable expression of enzymes essential for CMP-sialic acid regeneration while the genes located on the plasmid can be shuffled easily in accordance with specific needs. Similar strategy has been validated in synthesis of important biosynthetic precursor, shikimic acid for aromatic natural products *(48-50)*, we expect that new generation superbug will exploit a brand new way for large-scale synthesis of oligosaccharides.

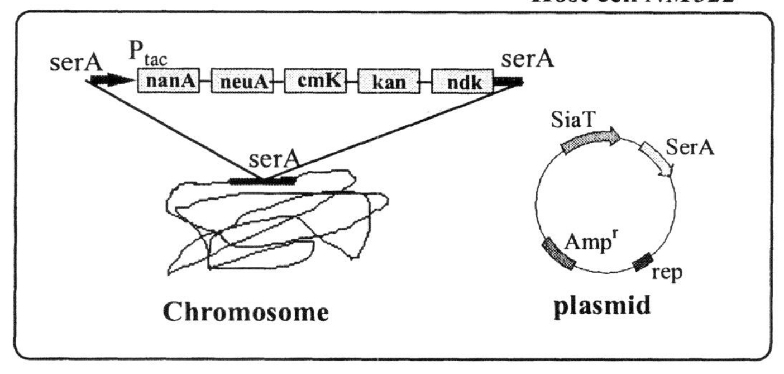

Fig 2 Construct of new generation of superbug

References

1. Nangia-Makker, P.; Conklin, J.; Hogan, V.; Raz, A. *Trends Mol Med* **2002,** *8*, 187-92.
2. Varki, A. *Glycobiology* **1993,** *3*, 97-130.
3. Kobata, A. *Eur J Biochem* **1992,** *209*, 483-501.
4. Fukuda, M. *Cancer Res* **1996,** *56*, 2237-44.
5. Tsuboi, S.; Fukuda, M. *Bioessays* **2001,** *23*, 46-53.
6. Palcic, M. M. *Current Opinion in Biotechnology* **1999,** *10*, 616-624.
7. Flitsch, S.L. *Current Opinion in Chemical Biology* **2000,** *4*, 619-625.
8. Depré, D.; Düffels, A.; Green, L. G.; Lenz, R.; Ley, S. V.; Wong, C. -H. *Chem. Eur. J.* **1999,** *5*, 3326-3340.
9. Ichikawa, Y.; Wang, R.; Wong, C. -H. *Methods in Enzymology* **1994,** *247*, 107-127.
10. Wong, C.-H.; Wang, R.; Ichikawa, Y. *J. Org. Chem.* **1992,** *57*, 4343-4344.
11. Chen, X.; Liu, Z.-Y.; Zhang, J.-B.; Zheng, W.; Kowal, P.; Wang, P. G. *ChemBioChem* **2002,** *3*, 47-53.
12. Chen, X.; Zhang, J.-B.; Kowal, P.; Liu, Z.-Y.; Andreana, P. R.; Lu, Y.; Wang, P. G. *J. Am. Chem. Soc.* **2001,** *123*, 8866-8867.
13. Yamada-Okabe, T.; Sakamori, Y.; Mio, T.; Yamada-Okabe, H. *Eur. J. Biochem.* **2001,** *268*, 2498–2505.
14. Hofmann, M.; Boles, E.; Zimmermann, F. K. *Eur J Biochem* **1994,** *221*, 741-747.

15. Brown, K.; Pompeo, F.; Dixon, S.; Mengin-Lecreulx, D.; Cambillau, C.; Bourne, Y. *EMBO J.* **1999,** *18*, 4096-4107.
16. Ponce, E.; Flores, N.; Martinez, A.; Valle, F.; Bolivar, F. *J. Bacteriol.* **1995,** *177*, 5719-5722.
17. Lahti, R.; Pitkaranta, T.; Valve, E.; Ilta, I.; Kukko-Kalske, E.; Heinonen, J. *J. Bacteriol.* **1988,** *170*, 5901–5907.
18. Shao, J.; Zhang, J. B.; Nahálkab, J.; Wang, P. G. *Chem. Comm.* **2002,** 2586-2587.
19. Kowal, P.; Wang, P. G. *Biochemistry* **2002,** *41*, 15410 - 15414.
20. Shao, J.; Zhang, J.; Kowal, P.; Wang, P. G. *Appl. Environ. Microbiol.* **2002,** *68*, 5634-5640.
21. Wu, B. –Y.; Zhang, Y.-X.; Zheng, R.; Guo, C.-W.; Wang, P. G. *FEBS letters* **2002,** *519*, 87-92.
22. Wu, B.-Y.; Zhang, Y.-X.; Wang, P. G. *Biochem. Biophys. Res. Commun.* **2001,** *285*, 364-371.
23. Ichikawa, Y.; Liu, J. L.-C.; Shen, G.-J.; Wong, C.-H. *J. Am. Chem. Soc.* **1991,** *113*, 6300-6302.
24. Ichikawa, Y.; Shen, G.-J.; Wong, C.-H. *J. Am. Chem. Soc.* **1991,** *113*, 4698-4700.
25. Zhang, J.-B.; Zhang, Y.-X.; Wu, B.-Y.; Wang, P. G. *Org. Lett.* **2003,** *in press.*
26. Simon, E. S.; Bednarski, M. D.; Whitesides, G. M. *J. Am. Chem. Soc.* **1988,** *110*, 7159-7163.
27. Huang, X.; Witte, K. L.; Bergbreiter, D. E.; Wong, C. -H. *Advanced Synthesis & Catalysis* **2001,** *343*, 675-681.
28. Toda, A.; Yamada, K.; Nishimura, S.-I. *Advanced Synthesis & Catalysis* **2002,** *344*, 61-69.
29. Nishiguchi, S.; Yamada, K.; Fuji Y.; Shibatani, S.; Toda A.; Nishimura, S.-I. *Chem. Commun.* **2001**, 1944–1945.
30. Yamada, K.; Nishimura, S.-I. *Tetrahedron Lett.* **1995,** *36*, 9493-9496.
31. Yan, F.; Wakarchuk, W. W.; Gilbert, M.; Richards, J. C.; Whitfield, D. M. *Carbohydr. Res.* **2000,** *328*, 3-16.
32. Yamada, K., Fujita, E., Nishimura, S.-I. *Carbohydr. Res.* **1997,** *305*, 443-461.
33. Chen, X.; Fang, J.; Zhang, J.; Liu, Z.-Y.; Andreana, P.; Kowal, P.; Wang, P. G. *J. Am. Chem. Soc.* **2001,** *123*, 2081-2082.
34. Endo, T.; Koizumi, S.; Tabata, K.; Kaita, S.; Ozaki, A. *Carbohydrate Research* **1999,** *136*, 179-183.
35. Koizumi, S.; Endo, T.; Tabata, K.; Ozaki, A. *Nat. Biotechnol.* **1998,** *16*, 847-850.
36. Endo, T., Koizumi, S., Tabata, K., Ozaki, A. *Appl Microbiol Biotechnol* **2000,** *53*, 257-261.

37. Endo, T., Koizumi, S., Tabata, K., Kakita, S., Ozaki, A. *Carbohydrate Research* **2001,** *330*, 439–443.
38. Koizumi, S., Endo, T., Tabata, K., Nagano, H., Ohnishi, J., Ozaki, A. *Journal of Industrial Microbiology & Biotechnology* **2000,** *25*, 213-217.
39. Fang, J.-W.; Li, J.; Chen, X.; Zhang, Y.-N.; Wang, J.-Q.; Guo, Z.-M.; Brew, K.; Wang, P. G. *J. Am. Chem. Soc.* **1998,** *120*, 6635-6638.
40. Fang, J.-W.; Chen, X.; Zhang, W.; Wang, J.-Q.; Andreana, P.; Wang, P.G. *J. Org. Chem.* **1999,** *64*, 4089-4094.
41. Liu, Z. -Y.; Zhang, J.-B.; Chen, X.; Wang, P. G. *ChemBioChem* **2002,** *3*, 348-355.
42. Zervosen, A.; Elling, L. *J. Am. Chem. Soc.* **1996,** *118*, 1836 - 1840.
43. Bulter, T.; Wandrey, C.; Elling, L. *Carbohydr. Res.* **1997,** *305*, 469-473.
44. Samain, E.; Drouillard, S.; Heyraud, A.; Driguez, H.; Geremia, R. A. *Carbohydrate Research* **1997,** *302*, 35-42.
45. Samain, E., Chazalet, Valerie., and Geremia, R.A. *Journal of Biotechnology* **1999,** *72*, 33-47.
46. Bettler, E.; Samain, E.; Chazalet, V.; Bosso, C.; Heyraud, A.; Joziasse, D. H.; Wakarchuk, W. W.; Imberty, A.; Geremia, R. A. *Glycoconj. J.* **1999,** *16*, 205-212.
47. Priem, B.; Gilbert, M.; Wakarchuk, W. W.; Heyraud, A.; Samain, E. *Glycobiology* **2002,** *12*, 235-240.
48. Ran, N.; Knop, D. R.; Draths, K. M.; Forst, J. W. *J. Am. Chem. Soc.* **2001,** *123*, 10927-10934.
49. Li, K.; Mikola, M. R.; Draths, K. M.; Worden, R. M.; Forst, J. W. *Biotechnology and Bioengineering* **1999,** *64*, 61-73.
50. Barker, J. L.; Forst, J. W. *Biotechnology and Bioengineering* **2001,** *76*, 376-390.

Indexes

Author Index

Blixt, Ola, 93
DeAngelis, Paul L., 125
Deng, Chao, 39
Eichler, Eva, 53
Gluzman, Ellen, 39
Ichikawa, Yoshi, 23
Jolly, Laure, 139
Kochhar, Sunil, 139
Koizumi, Satoshi, 153
Kowal, Przemyslaw, 1
Li, Hanfen, 1
Li, Mei, 1, 165
Mehta, Seema, 53
Monde, Kenji, 113
Nagahori, Noriko, 113
Ni, Jiahong, 73
Niikura, Kenichi, 113
Nishimura, Shin-Ichiro, 113
Plante, Obadiah J., 11
Rabuka, David, 23
Razi, Nahid, 93
Sadamoto, Reiko, 113
Sgarbi, Paulo W. M., 23
Shao, Jun, 1, 165
Singh, Suddham, 73
Tornare, Véronique, 139
Trummer, Brian J., 39
Wacowich-Sgarbi, Shirley A., 23
Wakarchuk, Warren, 53
Wang, Denong, 39
Wang, Lai-Xi, 73
Wang, Peng George, 1, 165
Wang, Ruby, 39
Whitfield, Dennis M., 53
Xiao, Min, 165
Yan, Fengyang, 53
Yi, Wen, 1

Subject Index

A

Acceptor-bound carbohydrate, synthesis, 14, 15*f*
N-Acetylheparosan, glycosaminoglycans, 126
Antibody recognition, carbohydrate structures, 44, 45*f*
Antigenic diversity, carbohydrates, 43–46, 49–50
Atomic force microscopy (AFM). *See* Glycosylation
Automated carbohydrate synthesis
  acceptor-bound, 14, 15*f*
  automation, 14
  cap-tag procedure for biopolymer purification, 20*f*
  drug discovery and development, 21
  future directions, 21
  glycosyl phosphates, 17, 18*f*
  glycosyl trichloroacetimidates, 16–17
  integrated donor method, 17, 19
  polymannosides, 16*f*
  strategy, 14
  synthesis using glycosyl trichloroacetimidates and glycosyl phosphates, 19*f*
  synthesis with glycosyl phosphates, 18*f*
  tagging and purification of compounds, 19–20
Automated glycosynthesizer. *See* Glycosylation

B

Bacteria. *See* Exopolysaccharides (EPSs); Genetically engineered bacteria
Bacterial capsules, glycosaminoglycans, 126
Bacterial coupling
  carbohydrate synthesis, 5–6
  future perspective, 162
  production of globotriose, 160*f*
  production of oligosaccharides, 159, 162
  production of 3'-sialyllactose, 161*f*
  production of sugar nucleotides, 157–158
  production of UDP-Gal, 158*f*
Bacterial glycosyltransferases, sources, 154–156
Biochip platform, high potential, 47–50
Biopolymers, purification, 19–20
Biosensor chip
  procedure to construct glycosyltransferase, 116
  See also Glycosylation
Biosignals, carbohydrates, 42–43
Biosynthesis pathway engineering
  cell free oligosaccharide synthesis, 171–172
  CMP-Neu5Ac synthesis, 169–171
  further improvements in metabolically engineered bacteria, 177–178
  GDP-mannose and GDP-fucose synthesis, 168–169

sugar nucleotide synthesis, 166–171
UDP-Glc and UDP-Gal synthesis, 166–167
UDP-GlcNAc and UDP-GalNAc synthesis, 167, 168
Wang's superbeads, 172, 173
Wang's superbug, 174–176
whole cell approaches for genetically engineered bacteria, 173–177
See also Sugar nucleotides
Blood types, specificity, 42–43

C

Calcitonin, glycosylation, 76–77
Cancer therapy, carbohydrate-based vaccine, 26, 27*f*, 28
Capping and tagging, biopolymers, 19–20
Capsule, polysaccharide, 126
Carbohydrates
antigenic diversity, 43–46, 49–50
biosignals, 42–43
components in drugs, 24
conformational flexibility, 49
covalent attachment, 2–3
current practices in production, 12–13
generating compound library, 95–96
medicinal chemistry by OPopS™ (optimer programmed one-pot synthesis), 36–37
purification factors, 95–96
roles in biological processes, 140
sialyltransferases for sialylated structures, 100
solution behavior, 44, 46
structural diversity, 12, 41
structures and antibody recognition, 44, 45*f*
traditional synthesis of trisaccharide, 24, 25*f*
See also Automated carbohydrate synthesis; OPopS™ (optimer programmed one-pot synthesis)
Carbohydrate synthesis, enzymatic
bacterial coupling technology, 5–6
glycosyltransferases, 4–5
living factory technology, 7–8
superbug technology, 6–7
Chemoenzymatic synthesis, sugar polymers, 126
Chemoenzymatic synthesis of lactosamine
β-linked trisaccharide from disaccharide donor, 60
chemical structure of repeat unit of group B *Streptococcus* Type 1A, 57
competition experiment between soluble disaccharide and polymer-supported, 63
2,5-dimethylpyrrole and 2,2',2"-trichloroethoxycarbonyl (Troc) protecting groups, 67–69
directed combinatorial chemistry, 70
future prospects, 69–70
ganglioside GM3 analogue, 62, 63
glycobiology, 53, 56
N-Troc protected sialylated lactosamine donor, 67
oligosaccharide linkages on soluble polymeric supports, 64
optimization of oligosaccharide linking, 54, 55
pentasaccharide repeat unit as target, 56, 60
program with *N*-phthaloyl protected building blocks, 66
protected versions of branched trisaccharide, 60–61
purification of products, 65–66
ring opening of *N*-phthalimide groups in water, 66
sialylated galactose disaccharide building block, 60
sialylated lactosamine with *N*-phthalimide as *N*-protecting group, 65
silalylated lactosamine derivative from disaccharide donor and

polymer-supported acceptor, 61
strategy for lactosamine or silalylated lactosamine building blocks, 61–63
tetrasaccharide representative of *N*-linked glycopeptide arm, 69
transformation of trisaccharide into pentasaccharide, 58, 59
trisaccharide from unreactive polymer-supported acceptor, 68
Chimeric synthases, novel glycosaminoglycans with, 132, 135
Chondroitin, glycosaminoglycans, 126
Cofactor regeneration, oligosaccharide synthesis, 156–157
Combinatorial chemistry, oligosaccharide building blocks, 70
Complex carbohydrates, biological information, 42–43
Conformational flexibility
carbohydrates, 49
sugar chains, 41
Conformational studies
glycoforms of complex glycopeptides, 80–81
See also Glycopeptides

D

Dextran
biosignals, 42
microarrays, 46–47, 48*f*
2,5-Dimethylpyrrole, protecting group, 67–68
Directed combinatorial chemistry, oligosaccharide building blocks, 70
Diversity. *See* Automated carbohydrate synthesis
Drug discovery
automated solid-phase synthesis, 21
OPopS™ (optimer programmed one-pot synthesis), 36–37

E

Endo-β-*N*-acetylglucosaminidases
transglycosylation activity of, 74–76
transglycosylation and hydrolysis, 75*f*
Endoglycosidase
synthesis of glycosylated bioactive peptides, 76–79
synthesis of substrate analogs for glycoamidases, 82–85
See also Glycopeptides
Engineered bacteria. *See* Exopolysaccharides (EPSs)
Enzymes
isolation of recombinant, 5
oligosaccharides and glycoconjugates, 4
transglycosylation activity of endo-β-*N*-acetylglucosaminidases, 74–76
See also Carbohydrate synthesis, enzymatic; Glycosaminoglycans (GAGs)
Escherichia coli, bacterial glycosyltransferases, 154–156
Escherichia coli O157, hamburger bug, 3
Eukaryotic proteins, glycosylation, 2–3
Exopolysaccharides (EPSs)
bacterial cells, 3
EPS gene cluster identification, 143, 144*f*
fucosylation of EPS, 149*f*
genetic organization of EPS gene clusters from *S. macedonicus*, 144*f*
glycosyltransferases (GTFs), 140, 143, 145
in vitro modification, 148, 150
in vivo EPS engineering, 145, 148
in vivo modification, 141, 142*f*
knocking out gene coding for GTF modifying, 148
lactic acid bacteria (LAB) for production, 140–141

putative functions of GTFs from *S. macedonicus* Sc136, 145, 147*f*
recombinant GTFs, 140
structure of EPS from *S. macedonicus* Sc136, 142*f*
thin layer chromatography (TLC) analysis of activity assays with purified GTFs, 145, 146*f*

F

Flexibility
carbohydrates, 48
sugar chains, 41
Fucosylation
exopolysaccharides, 149*f*
in vitro modification of exopolysaccharides, 148, 150
Fucosyltransferases (FUTs), galactosides and polylactosamines, 97, 99*f*

G

Galactosides
fucosyltransferases, 97, 99*f*
synthesis, 96–97
Ganglioside, synthesis of mimics, 106, 108*f*, 109*f*
Genetically engineered bacteria
bacterial glycosyltransferases, 154–156
future perspective, 162
improvements in metabolically engineered bacteria system, 177–178
living cell synthesis oligosaccharide, 177
oligosaccharide synthesis with cofactor regeneration, 156–157
production of globotriose by bacterial coupling, 160*f*
production of oligosaccharides by bacterial coupling, 159, 162
production of 3'-sialyllactose by bacterial coupling, 161*f*
production of sugar nucleotides by bacterial coupling, 157–158, 159*f*
recycling systems, 156–157
sugar nucleotide recycling by sucrose synthetase, 157*f*
technology through multiple bacterial coupling, 174
Wang's superbug, 174–177
whole cell process, 173–174
Genome-sequencing projects, biological and medical research, 40–41
Globo H hexasaccharide
synthesis, 28, 30*f*
See also OPopS™ (optimer programmed one-pot synthesis)
Globotriose
bacterial coupling, 160*f*
synthesis of, and derivatives, 175–176
Glycoamidases, synthesis of substrate analogs for, 82–85
Glycobiology
industrialization, 53, 56
research area, 165
Glycoconjugates
biochip platform, 47, 49–50
enzymatic synthesis, 4
Glycolipids
general composition, 3
synthesis of ganglioside mimics, 106, 108*f*, 109*f*
Glycomers, term, 40
Glycomics, scientific discipline, 40
Glycopeptides
glycoforms of complex, for conformational studies, 80–81
glycoforms of extracellular loop peptide from nicotinic acetylcholine receptor, 81*f*
glycosylated α-mating factor and peptide T, 79, 80*f*
glycosylated calcitonin, 76–77
glycosylated substance P, 78–79
structures of high-mannose type gp120, 86*f*
structures of substrate analogs for glycoamidase, 82*f*

synthesis of double-glycosylated gp120, 87
synthesis of human immunodeficiency virus type 1 (HIV-1) envelope glycoprotein gp120, 85–88
synthesis of substrate analogs for glycoamidases, 82–85
transglycosylation activity of endo-β-*N*-acetylglucosaminidases, 74–76
Glycoproteins, microarray, 47, 48*f*
Glycosaminoglycans (GAGs)
analysis of chimeric synthases, 135*f*
bacterial capsules and, 126
composition, 126
construction of oligosaccharide libraries, 132
domain structures of *Pasteurella* GAG synthases, 130
elongation activity of recombinant *Pasteurella* synthases, 128–129
GAG glycosyltransferases (GAG synthases), 127
hybrid GAG polymers, 130, 131*f*
MALDI–TOF MS analysis of fluorescent hyaluronan (HA) oligosaccharide, 133*f*
model of reaction mechanism, 129*f*
novel GAGs with chimeric synthases, 132, 135
Pasteurella GAG and synthases, 127*t*
Pasteurella GAG synthases, 127–128
potential biomedical agents using novel amino-GAGs, 136*f*
schematic of creation and use of catalysts for oligosaccharide synthesis, 133*f*
schematic of hyaluronan synthase (pmHAS) and chondroitin synthase (pmCS) domain structure, 129*f*
schematic of synthesis of variant GAG sugar, 134*f*
screening of GAG oligosaccharide library, 134*f*
structural complexity, 12
sugar specificity of chimeric enzyme pmCHC, 136*t*
sugar transfer specificity, 127
synthesis of defined, monodisperse oligosaccharides, 130, 132
Glycosidases, glycan processing reactions, 4
Glycosylation
atomic force microscopy (AFM) images of maltose-binding protein-galactosyltransferase (MBP-GalT) on glycolipid LB films, 118*f*
cell type and tissue type dependence, 2
chemical structures of photopolymerizable lipids, 116*f*
concept of engineered glycosyltransferases by fusion protein with MBP, 114, 115*f*
effect of sugar density of LB film on activity of MBP-GalT on sensor chip, 119, 121*f*
eukaryotic proteins, 2–3
evaluation of immobilized MBP-GalT on gold sensor chips, 118–119
experimental, 122–123
fusion proteins with MBP, 114
monitoring substrate binding to MBP-GalT array on SPR sensor chip, 123
morphology of MBP-GalT bound on surface of Langmuir–Blodgett (LB) films by AFM, 118
polydiacetylene-type glycolipid LB membrane, 115
preparation of GalT microarray, 122–123
procedure to construct glycosyltransferase biosensorchip, 116
process of galactosylation on surface plasmon resonance (SPR) sensor chip, 120*f*
protein, 43
protein modification, 74
real-time monitoring, 114

selected spectral data for product, 123
SPR sensorgram on adding glycosyl donor and/or acceptor substrates, 120*f*
synthesis of photopolymerizable glycerolglycolipid, 122
synthesis of photopolymerizable glycolipid, 117
See also Glycopeptides
Glycosyl phosphates, automated solid-phase synthesis, 17, 18*f*, 19*f*
Glycosylphosphatidyl inositol molecules (GPIs), structural complexity, 12
Glycosyltransferases
availability of new, 5
bacterial, 154–156
biosynthesis of oligosaccharide by, 4
identification and cloning, 5
oligosaccharide synthesis, 140
specificity, 5, 13
Glycosyl trichloroacetimidates, automated solid-phase synthesis, 16–17, 19*f*
Gold sensor chips, immobilized maltose binding protein-galactosyltransferase on, 118–119

H

Hamburger bug, *Escherichia coli* O157, 3
Heparosan, glycosaminoglycans, 126
Human immunodeficiency virus type-1 (HIV-1), synthesis of HIV-1 envelope glycoprotein gp120 glycopeptides, 85–88
Human milk oligosaccharide (HMO), lacto-*N*-neotetraose (LNnT), 140
Hyaluronan (HA), glycosaminoglycans, 126
Hydrolysis, endo-β-*N*-acetylglucosaminidases, 75*f*

I

Industrialization, glycobiology, 53, 56
Infection strategy, *Escherichia coli* O157, 3

L

Lactic acid bacteria
exopolysaccharide production, 140–141
See also Exopolysaccharides (EPSs)
Lacto-*N*-neotetraose (LNnT), human milk oligosaccharide, 140
Lactosamine. *See* Chemoenzymatic synthesis of lactosamine
Langmuir–Blodgett (LB) films
polydiacetylene-type glycolipid LB membrane, 115
See also Glycosylation
Libraries
construction of oligosaccharide, 132, 134*f*
generating carbohydrate compound, 95–96
Lipooligosaccharide (LOS), bacterial cells, 3
Lipopolysaccharide (LPS), bacterial cells, 3
Living cell synthesis, oligosaccharide, 177
Living factory technology, carbohydrate synthesis, 7–8

M

Maltose-binding protein (MBP)
cell free oligosaccharide synthesis, 171–172
glycosyltransferases as fusion proteins with MBP, 114, 115*f*

See also Glycosylation
α-Mating factor
glycosylation, 79, 80*f*
See also Glycopeptides
Medicinal chemistry, OPopS™ (optimer programmed one-pot synthesis), 36–37
Microarrays
antigen, for detection of human and murine antibodies, 48*f*
biological and medical research, 40–41
carbohydrate-based technology, 46-47
dextran, 46–47
high potential biochip platform, 47–50
photopolymerizable glycolipids for carbohydrate-based, 115
Microorganisms, sugar chain recognition, 43
Modification. *See* Exopolysaccharides (EPSs)
Monitoring. *See* Glycosylation

N

N-acetylheparosan, glycosaminoglycans, 126
Neuropeptide substance P
glycosylation, 78–79
See also Glycopeptides
N-phthalimide protecting group, ring opening in water, 66

O

Oligosaccharides
bacterial coupling, 159, 162
biosynthesis by glycosyltransferase, 4
cell free synthesis, 171–173
chemical synthesis, 165
chemoenzymatic iterative synthesis of linkages, 64
chemoenzymatic synthesis strategy, 94–95
components or determinants of antigen, 43
construction of libraries, 132, 134*f*
creation and use of catalysts for synthesis, 133*f*
directed combinatorial chemistry for building blocks, 70
enzymatic activity of α-D-galacto derivatives, 102, 103*t*
enzymatic one-pot synthesis of sialyloligosaccharides, 101*f*
enzymatic preparation of O-linked sialosides, 104*f*, 105*f*
enzymatic synthesis, 4, 165–166
future perspective in production, 162
large-scale production through bacterial coupling, 6
living cell synthesis, 177
maltose-binding protein (MBP), 171–172
optimization of linking, 54, 55
potential pharmaceuticals, 3
purification, 139
recognition markers, 2
synthesis of defined monodisperse, 130, 132
synthesis of sialic-acid-containing, 100, 102
synthesis with cofactor regeneration, 156–157
Wang's superbeads, 172
Wong's and Nishimura's method, 171–172
See also Bacterial coupling; Chemoenzymatic synthesis of lactosamine; Exopolysaccharides (EPSs); OPopS™ (optimer programmed one-pot synthesis)
OPopS™ (optimer programmed one-pot synthesis)
cancer antigens expression, 29*f*
carbohydrate-based cancer vaccine, 26, 28
deprotection of Globo H hexasaccharide, 34*f*

Globo H hexasaccharide synthesis, 28, 30*f*
Globo H synthesis, 33*f*, 34*f*
key elements, 24, 26
medicinal chemistry, 36–37
novel cancer therapy, 27*f*
one-pot approach II with sugar building blocks with α-galactosyl linkage, 32–36
one-pot approach I with three glycosylations, 28, 31–32
one-pot trisaccharide synthesis using OPopS™, 27*f*
relative reactivity values (RRVs), 24, 26
synthesis of disaccharide building block, 31*f*
synthesis of Globo H tetrasaccharide, 35*f*

P

Pasteurella
domain structures of glycosaminoglycan synthases, 130
elongation activity of recombinant, synthases, 128–129
glycosaminoglycan synthases, 127–128
See also Glycosaminoglycans (GAGs)
Peptide T
glycosylation, 79, 80*f*
See also Glycopeptides
Pharmaceuticals, potential oligosaccharides, 3
Phosphorylation, protein, 43
Photopolymerizable lipids
chemical structures, 116*f*
synthesis, 117
synthetic procedure, 122
See also Glycosylation
N-Phthalimide protecting group, ring opening in water, 66
Plasticity, sugar chains, 41
Polymannosides, automated solid-phase synthesis, 16–17
Poly-*N*-acetyllactosamines
fucosyltransferases, 97, 99*f*
synthesis, 96–97
Polysaccharides
complete antigens, 43
synthesis with enzymes, 2
See also Exopolysaccharides (EPSs); OPopS™ (optimer programmed one-pot synthesis)
Protecting groups. *See* Chemoenzymatic synthesis of lactosamine
Protein
glycosylation, 43, 74
See also Glycopeptides
Purification
biopolymers, 19–20
enzymatic synthesis of carbohydrates, 95–96

R

Real-time monitoring. *See* Glycosylation
Recognition markers, oligosaccharides, 2
Recombinant enzymes
elongation activity of recombinant *Pasteurella* synthases, 128–129
isolation, 5
living cell synthesis oligosaccharide, 177
sialyltransferases for oligosaccharides, 100, 102
Recycling, sugar nucleotide systems, 156–157
Regeneration, cofactor, oligosaccharide synthesis, 156–157
Relative reactivity values (RRVs), OPopS™ technology, 24, 26

S

Sialic acid, common derivatives, 100
Sialosides, enzymatic preparation of O-linked, 104*f*, 105*f*

Sialylated lactosamine. *See* Chemoenzymatic synthesis of lactosamine
3'-Sialyllactose, bacterial coupling, 161*f*
Sialyltransferases, sialylated carbohydrate structures, 100
Solution behavior, carbohydrates, 44, 46
Specificity
- blood types, 42–43
- glycosyltransferases, 5
- poly-*N*-acetyllactosamines, 96–97, 98*t*
- sugar, of enzyme pmCHC, 136*t*
- sugar transfer, 127

Structural diversity, carbohydrates, 12, 41
Substance P neuropeptide
- glycosylation, 78–79
- *See also* Glycopeptides

Sugar building blocks, OPopS™ technology, 24
Sugar engineering. *See* Glycosaminoglycans (GAGs)
Sugar nucleotides
- CMP-Neu5Ac regeneration cycle, 170
- GDP-Fuc regeneration cycle, 170
- GDP-Man regeneration cycle, 169
- production by bacterial coupling, 157–158
- synthesis of CMP-Neu5Ac, 169–171
- synthesis of GDP-mannose and GDP-fucose, 168–169
- synthesis of globotetraose, 167, 168
- synthesis of UDP-Glc and UDP-Gal, 166–167
- UDP-GclNAc and UDP-GalNAc, 167

Sugar polymers, chemoenzymatic synthesis, 126
Sugar transfer, specificity, 127
Superbeads, cell free oligosaccharide synthesis, 172, 173
Superbug technology
- carbohydrate synthesis, 6–7
- construct of new generation of, 179*f*
- development of new generation, 178
- strategy, 175
- Wang's, 174–176

Surface plasmon resonance (SPR). *See* Glycosylation
Synthesis. *See* Automated carbohydrate synthesis

T

Tagging, biopolymers, 19–20
Transglycosylation
- activity of endo-β-*N*-acetylglucosaminidases, 74–76
- *See also* Glycopeptides

2,2'2"-Trichloroethoxycarbonyl (Troc), protecting group, 67–69
Trisaccharide, traditional synthesis, 24, 25*f*

W

Whole cell approach
- further improvements, 177–178
- living cell synthesis oligosaccharide, 177
- synthesis of α-gal epitope, 175
- synthesis of globotriose and derivatives, 175–176
- Wang's superbug, 174–17